Civil Engineering PE Practice Exams: 2 Full Breadth Exams

By Civil PE Practice
2nd Printing

Register This Book and get additional practice questions

https://civilpepractice.com

© 2018 Civil PE Practice, LLC. All rights reserved.
All content is copyrighted by Civil PE Practice. No part, either text or image, may be used for any purpose other than personal use. Reproduction, modification, storage in a retrieval system or retransmission, in any form or by any means, electronic, mechanical, or otherwise, for reasons other than personal use is strictly prohibited.

The author and the publisher do not warrant that the information contained in this book is fully complete and shall not be responsible for any errors or omissions. The author and publisher shall have neither liability nor responsibility to any person or entity with respect to any damage related directly or indirectly to this book. All individuals shall consult applicable building or engineering codes prior to performing any engineering related work.

ISBN 978-1983913686

Breadth Exam

Version A

AM PRACTICE EXAM – Version A

1. A +2.25% grade intersects a -2.00% grade at Sta. 25+00 and elevation 105.75 ft. A 1000 ft vertical curve connects the two grades. The elevation along the curve at Sta. 24+25 is most nearly:

   (A) 94.50
   (B) 99.68
   (C) 100.22
   (D) 101.46

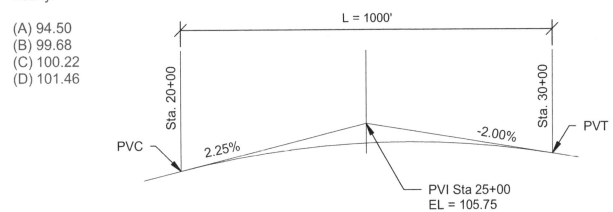

AM PRACTICE EXAM – Version A

2. A building floor plan is shown below. A 12-inch thick normal weight concrete floor is supported by concrete beams and columns as shown. The floor supports a superimposed live load of 100 psf. The maximum unfactored moment (k-ft) in beam B-1 is most nearly:

Note: Ignore self-weight of concrete beams and columns. Assume all beams are simply supported. The unit weight of concrete = 150pcf.

(A) 47.2
(B) 49.6
(C) 50.7
(D) 53.4

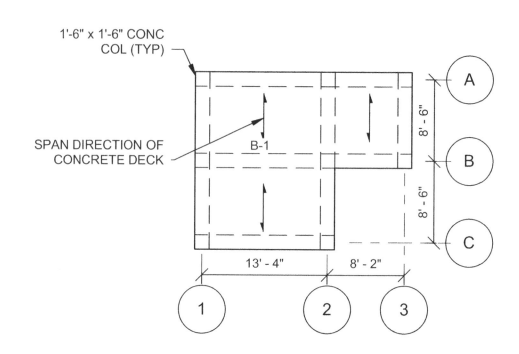

AM PRACTICE EXAM – Version A

3. A 30-ft thick clay soil layer is shown below. The clay has a coefficient of consolidation of 0.26 ft²/day. The time (days) when 40% of the total settlement will occur is most nearly:

   (A) 109
   (B) 218
   (C) 436
   (D) 475

U = degree of consolidation
T_v = time factor

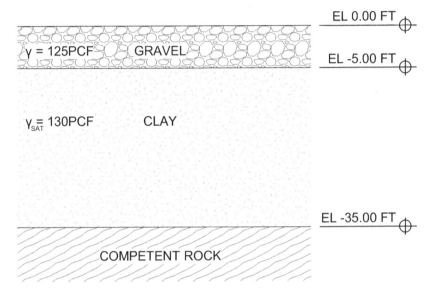

U	Tv
0.10	0.008
0.20	0.031
0.30	0.071
0.40	0.126
0.50	0.197
0.55	0.238
0.60	0.287
0.65	0.340
0.70	0.403
0.75	0.477
0.80	0.567
0.85	0.684
0.90	0.848
0.95	1.129
0.99	1.781

4. The maximum shear force (kips) for the beam shown below is most nearly:

   (A) 9.0
   (B) 15.0
   (C) 18.0
   (D) 25.0

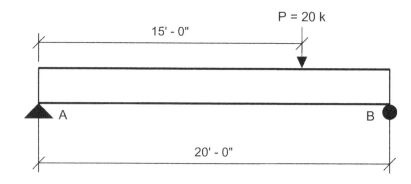

5. Fatigue of structural steel is best defined as:

   (A) Deflection exceeding recommended limits
   (B) Excessive shear deformation
   (C) Weakening caused by repeatedly applied loads
   (D) Failure due to excessive bending stress

AM PRACTICE EXAM – Version A

6. The effective vertical stress (psf) at Point A of the soil profile shown is most nearly:

   (A) 2265
   (B) 3265
   (C) 3555
   (D) 3825

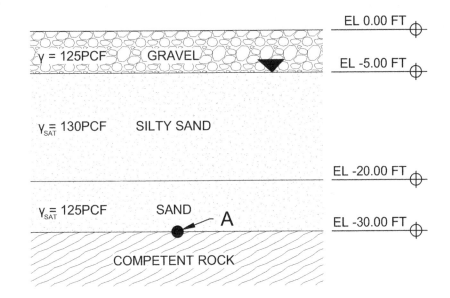

7. A family purchases a vacation home adjacent to a large lake. The family plans on using the vacation home every summer for the next 25 years. The probability that a 100-year flood does not occur over the next 25 years is most nearly:

   (A) 0.222
   (B) 0.546
   (C) 0.778
   (D) 0.990

AM PRACTICE EXAM – Version A

8. A vertical excavation is required at the construction site of a new office building. An 8 ft deep vertical cut is made in deep clay soils. The clay has a unit weight and an unconfined compressive strength of 120 pcf and 2000 psf respectively. The factor of safety of the vertical cut is most nearly:

   (A) 2.08
   (B) 2.25
   (C) 2.35
   (D) 2.45

9. Speed data was collected as part of a transportation study of a rural town. The data was collected on a moderately traveled two lane collector road. The speed data is given below. The time mean speed (mph) is most nearly:

   (A) 50.00
   (B) 50.15
   (C) 50.23
   (D) 50.36

Recorded Speed (mph)	Frequency
46	16
47	30
48	67
49	105
50	155
51	112
52	84
53	46
54	18

Copyright 2018 by CivilPEPractice
For more problems visit https://civilpepractice.com/

10. The depth of flowing water for the given channel is 3.5 ft. The velocity (fps) of the water is most nearly:

Given:
Manning's roughness = 0.025
Channel slope = 1.0%

(A) 9.1
(B) 9.5
(C) 10.4
(D) 13.9

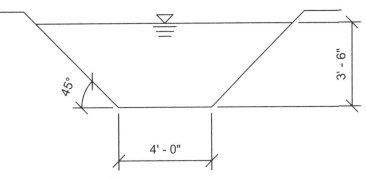

11. The ground elevation at point C is most nearly:

    (A) 104.30
    (B) 105.55
    (C) 106.12
    (D) 106.79

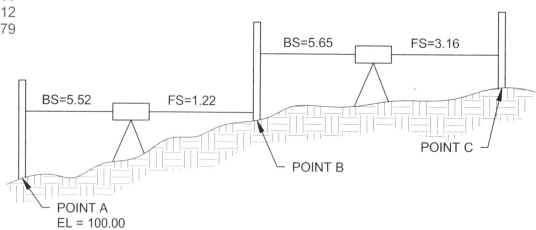

12. A 12-ft thick soil layer has the following properties:

$C_c = 0.14$ $C_r = 0.02$
$e_o = 0.92$ $\sigma'_o = 1,700 \, psf$
$\sigma'_c = 2,300 \, psf$

An additional 2,000 psf vertical load will be applied to the top of the soil layer. The primary consolidation (in) is most nearly:

(A) 0.20
(B) 0.50
(C) 1.00
(D) 2.40

13. The properties for water flowing through a 6-inch (I.D.) closed conduit are given below. The flow is best classified as:

    (A) Laminar
    (B) Transitional
    (C) Turbulent
    (D) None of the above

    Given:

    $$\text{velocity} = 0.5 \, \frac{\text{ft}}{\text{sec}}$$

    $$\text{kinematic viscosity} = 1 \cdot 10^{-5} \, \frac{\text{ft}^2}{\text{sec}}$$

14. The floor framing shown below supports the following loads:

    6 in normal weight (150 pcf) reinforced concrete slab
    suspended ceiling = 3 psf
    mechanical piping = 5 psf

    The unfactored superimposed dead load (k/ft) on member A is most nearly:

    (A) 0.52
    (B) 0.65
    (C) 0.72
    (D) 1.20

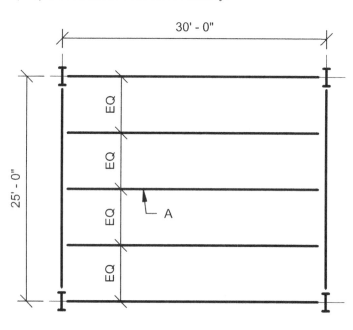

15. Which of the following work-related incidents must be reported to the Occupational Safety and Health Administration (OSHA)?

(A) Fatality
(B) Fatality and loss of eye
(C) Fatality, amputations, and loss of eye
(D) Fatality, broken arm resulting in hospitalization, amputations, and loss of eye

16. A steel pipe is reduced from 1' – 6" diameter to 6" diameter as shown below. The flow rate through pipe #1 is 5 cfs. The velocity (ft/sec) through pipe #3 is most nearly:

(A) 21.1
(B) 21.6
(C) 22.4
(D) 25.5

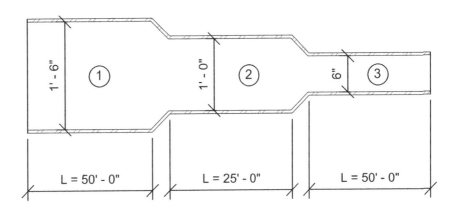

17. A concrete pipe encasement is constructed around a 42 in diameter (O.D.) steel pipe. The steel pipe connects two manholes which are 125 ft apart. The volume of concrete (yd³) required to encase the pipe is most nearly:

   (A) 15.4
   (B) 71.2
   (C) 115.7
   (D) 1922.4

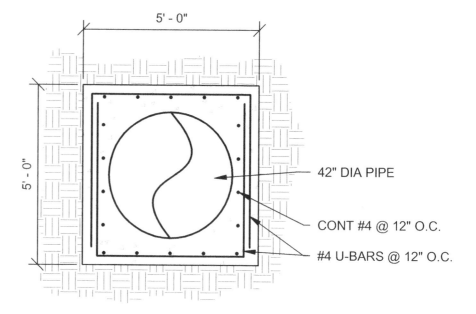

18. A 0.50 ft³ soil sample weighs 65.2 lb. The sample has a moisture content of 0.17 and specific gravity of 2.75. The degree of saturation is most nearly:

   (A) 0.54
   (B) 0.65
   (C) 0.75
   (D) 0.87

19. A six-story apartment complex is being construction adjacent to a large residential subdivision. Soil borings show 15ft of poor soils over a thick, competent rock layer. Which of the following foundation systems is best suited for the application?

   (A) Spread footings
   (B) Driven piles
   (C) Mat foundation
   (D) Drilled piles

20. The total float (days) for activity G is most nearly:

   Note: all duration values in the table below are given as days.

   (A) 0
   (B) 1
   (C) 2
   (D) 3

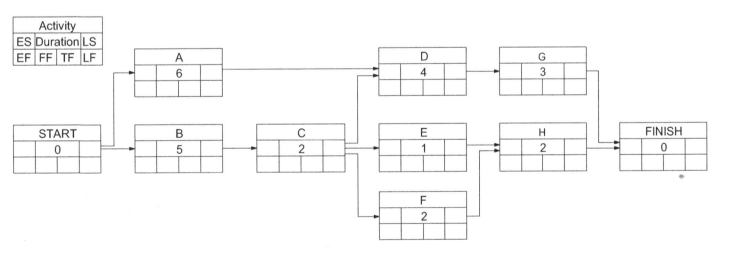

AM PRACTICE EXAM – Version A

21. A contractor is preparing to construct a 10,000 square foot steel framed building. Prior to pouring the slab-on-grade the contractor plans to compact the subgrade which is comprised of plastic clays. Which of the following compactors is best suited for the site?

    (A) Sheepsfoot Roller
    (B) Vibratory Plate Compactor
    (C) Vibrating Smooth Wheel Roller
    (D) Pneumatic Tire Roller

22. A 12,000 lb beam is lifted with the rig shown below. The tension force (lb) in Sling B is most nearly:
    Note: assume weight of the beam is uniformly distributed.

    (A) 6000
    (B) 6200
    (C) 6700
    (D) 6800

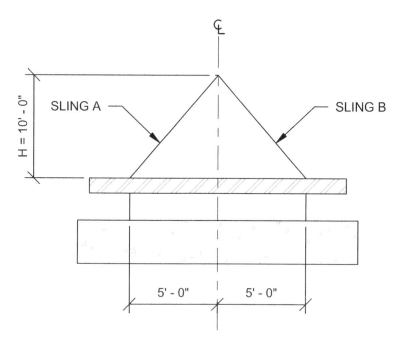

23. Which of the following is utilized as a sediment control method?

    (A) Geotextiles
    (B) Straw mulch
    (C) Silt fence
    (D) Hydraulic mulch

24. The contractor selects an excavator with a bucket capacity of 4-yd³ and a standard operating cycle time of 50 seconds for his project. The number of excavators required to excavate 7500 yd³ of soil in 8 hours is most nearly:

    (A) 3
    (B) 4
    (C) 5
    (D) 6

25. The maximum bearing pressure (psf) for the footing shown below is most nearly:

Note: all loads are unfactored.

(A) 5775
(B) 5850
(C) 5925
(D) 6250

Given:
B = 3.5 ft
L = 4 ft
P = 40,000 lb
M = 25,000 lb-ft

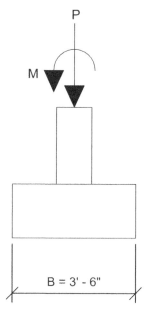

26. Which of the following diagrams best matches the activity data found in the table below?

(A) Diagram A
(B) Diagram B
(C) Diagram C
(D) Diagram D

Activity	Successor	Duration (days)
Start (A)	B,C	
B	D	6
C	D,E	8
D	F	4
E	F	3
F	G,H	4
G	J	6
H	J	7
Finish (J)		

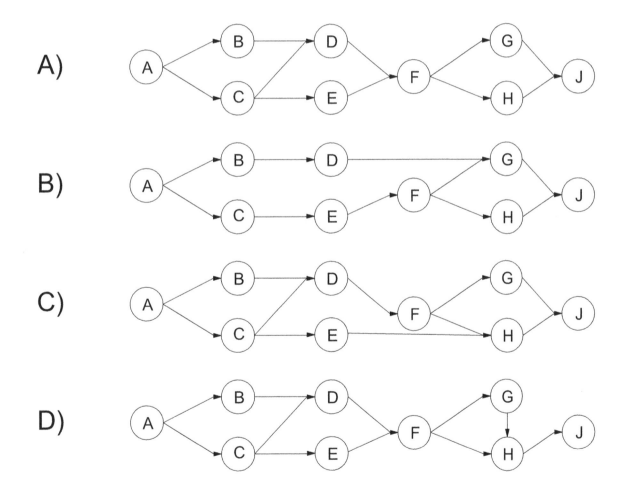

27. The slenderness ratio for the steel column shown below is most nearly:

(A) 8.16
(B) 48.98
(C) 73.47
(D) 97.96

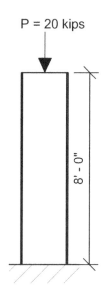

Column Properties	
Weight	50 plf
Area	14.7 in²
Depth	12 in
Radius of gyration	1.96 in

28. The length of tangent (feet) for a horizontal simple circular curve with a radius equal to 2500 feet and an intersection angle of 80 degrees is most nearly:

(A) 1000
(B) 1500
(C) 2100
(D) 2800

29. A continuous 3 ft wide footing bears on a thick layer of clean sand. The bottom of footing is located 4 ft below grade. Use Terzaghi's method to determine the maximum net allowable bearing capacity (psf):

(A) 3500
(B) 4500
(C) 5500
(D) 6500

Given:

$$\gamma = 125 \frac{lb}{ft^3} \qquad \phi = 30 \qquad FS = 3.0$$

$$N_c = 37.16 \qquad N_q = 22.46 \qquad N_\gamma = 19.73$$

30. The force (kips) in member CH of the truss shown below is most nearly:

(A) 18.4 Tension
(B) 18.4 Compression
(C) 20.0 Tension
(D) 20.0 Compression

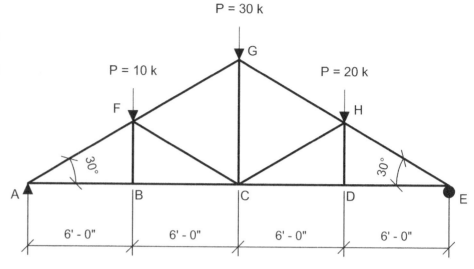

31. A cylindrical tank has a diameter of 18ft and is 16ft deep. The flow rate through the tank is 200 gpm. The detention time (hrs) is most nearly:

(A) 2.54
(B) 2.65
(C) 116.52
(D) 152.28

32. The water/cement ratio for the following concrete mix is most nearly:

(A) 0.53
(B) 0.50
(C) 0.45
(D) 0.40

Material	Material Type	ASTM	Weight (lbs)
Cement	Type I-II	C 150	490
Fly Ash	Fly Ash F	C - 618	95
Coarse Aggregate	# 57	C 33	1724
Fine Aggregate	Natural Sand	C 33	1285
Water	City		262
Admixture	Air Entrainer	C 260	3.5 oz/CY
Admixture	Type A Water Reducer	C 260	4.0 oz/CY

33. The total material cost of concrete for the tank shown below is most nearly:

Note: Assume a concrete material unit cost of $200 per cubic yard.

(A) $25,600
(B) $21,400
(C) $20,400
(D) $18,400

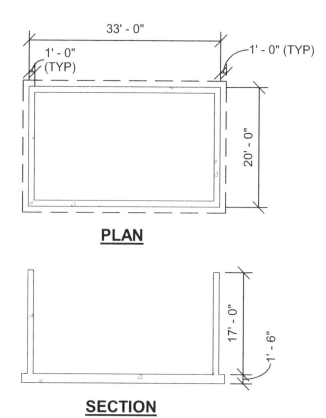

AM PRACTICE EXAM – Version A

34. A 500' long trench is to be cut to install steel piping. The area to be excavated is shown below. The total volume (yd³) of the excavation is most nearly:

   (A) 1900
   (B) 2000
   (C) 2100
   (D) 2200

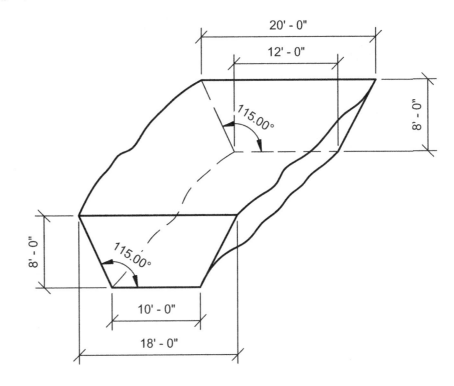

35. A concrete gravity wall with flat backfill retains 12' of sand. The active resultant force (lb/ft) induced on the wall due to lateral earth pressure is most nearly:

(A) 374
(B) 775
(C) 2244
(D) 2525

Soil Properties:

$\gamma_{soil} = 115 \dfrac{lb}{ft^3}$

$\phi = 35$

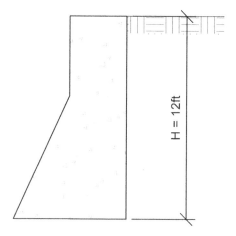

36. The contractor prepares the area to perform a concrete slump test. Which of the following is an acceptable surface in which to conduct the test?

(A) 5/8" solid sheet of plywood
(B) 4ft x 4ft level gravel surface
(C) 1/8" thick steel plate
(D) All of the above

AM PRACTICE EXAM – Version A

37. The USCS classification of the following soil sample is:

    (A) SC
    (B) SM
    (C) SP
    (D) GC

Sieve Size	Percent Finer
No. 4	90
No. 10	80
No. 40	70
No. 200	40

PL	16
LL	42

38. A 6-inch diameter (I.D.) pipe is used for a 5,500 ft segment of a water distribution system. The design velocity and friction factor are 6.5 ft/s and 0.0160, respectively. The friction loss (ft) of the pipe segment is most nearly:

    (A) 102
    (B) 107
    (C) 115
    (D) 126

39. The axial force (lb) resisted by a single temporary brace is most nearly:

Note: assume wall base is pinned and the brace only resists lateral load.

(A) 1745
(B) 2468
(C) 2577
(D) 2596

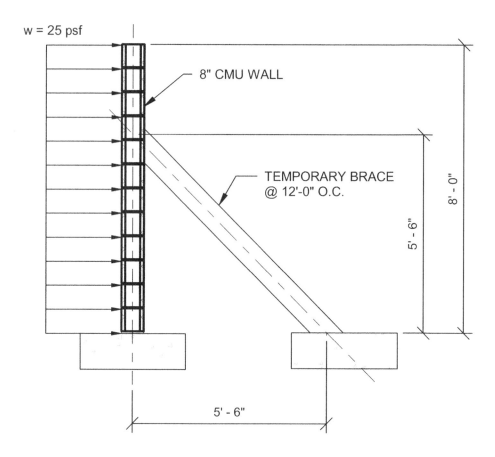

40. A catch basin is to be installed adjacent to a 3.4 acre parking lot. The design rainfall intensity is 6.7 in/hr and the runoff coefficient is 0.84. The peak discharge (ft³/sec) to the catch basin is most nearly:

(A) 15.4
(B) 16.3
(C) 16.7
(D) 19.1

Breadth Exam

Version A

Answers

AM PRACTICE EXAM – Version A ANSWERS

1. A +2.25% grade intersects a -2.00% grade at Sta. 25+00 and elevation 105.75 ft. A 1000 ft vertical curve connects the two grades. The elevation along the curve at Sta. 24+25 is most nearly:

   (A) 94.50
   (B) 99.68
   (C) 100.22
   (D) 101.46

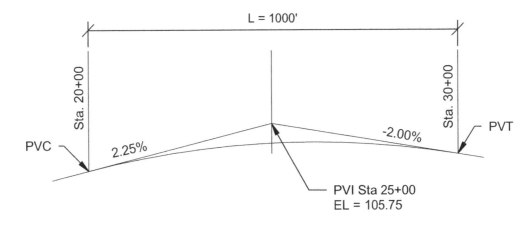

Given:

$G_1 = 0.0225$

$G_2 = -0.020$

$L = 1000$ ft

$PVI_{EL} = 105.75 ft$

Solve for elevation at Sta. 24+25:

$$R = \frac{(G_2 - G_1)}{L} = \frac{(-0.02 - 0.0225)}{1000\,ft} = \frac{-4.25 \cdot 10^{-5}}{ft}$$

$$x = (Sta.24 + 25) - Sta_{PVC} = 2425\,ft - 2000\,ft = 425\,ft$$

$$PVC_{EL} = PVI_{EL} - G_1 \cdot \left(\frac{L}{2}\right) = 105.75\,ft - 0.0225 \cdot \left(\frac{1000\,ft}{2}\right) = 94.50\,ft$$

$$EL_{24+25} = \frac{R}{2} \cdot x^2 + G_1 \cdot x + PVC_{EL}$$

$$EL_{24+25} = \frac{\frac{-4.25 \cdot 10^{-5}}{ft}}{2} \cdot (425\,ft)^2 + 0.0225 \cdot 425\,ft + 94.50\,ft = 100.224\,ft$$

Answer: 100.22

AM PRACTICE EXAM – Version A ANSWERS

2. A building floor plan is shown below. A 12-inch thick normal weight concrete floor is supported by concrete beams and columns as shown. The floor supports a superimposed live load of 100 psf. The maximum unfactored moment (k-ft) in beam B-1 is most nearly:

Note: Ignore self-weight of concrete beams and columns. Assume all beams are simply supported. The unit weight of concrete = 150pcf.

(A) 47.2
(B) 49.6
(C) 50.7
(D) 53.4

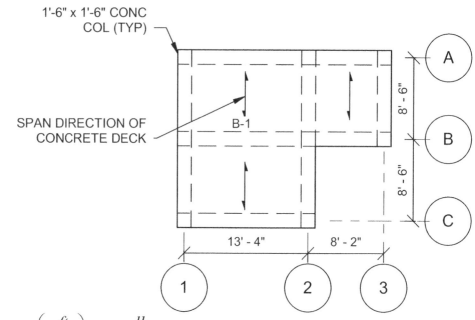

Given:

$$LL = 100 \frac{lb}{ft^2}$$

$$L = 13.333\, ft$$

$$DL = 150 \frac{lb}{ft^3} \cdot (12 in) \cdot \left(\frac{ft}{12 in}\right) = 150 \frac{lb}{ft^2}$$

Calculate tributary width of concrete beam:

$$A_{trib} = \frac{8.5\, ft}{2} + \frac{8.5\, ft}{2} = 8.5\, ft$$

Calculate load supported by concrete beam B-1:

$$w_{tot} = (DL + LL) \cdot A_{trib} = 2{,}125 \frac{lb}{ft} = 2.125 \frac{k}{ft}$$

Solve for maximum moment of beam (simply supported beam):

$$M_{max} = \frac{w_{tot} \cdot L^2}{8} = \frac{\left(2.125 \frac{k}{ft}\right) \cdot (13.333\, ft)^2}{8} = 47.222\, k \cdot ft$$

Answer: 47.2

AM PRACTICE EXAM – Version A ANSWERS

3. A 30-ft thick clay soil layer is shown below. The clay has a coefficient of consolidation of 0.26 ft²/day. The time (days) when 40% of the total settlement will occur is most nearly:

   (A) 109
   (B) 218
   (C) 436
   (D) 475

U=degree of consolidation
T_v=time factor

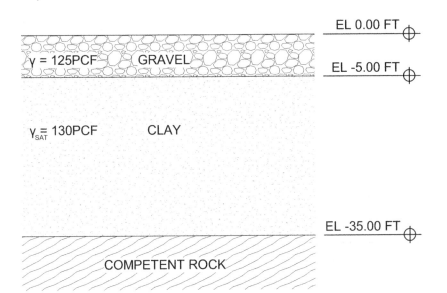

U	Tv
0.10	0.008
0.20	0.031
0.30	0.071
0.40	0.126
0.50	0.197
0.55	0.238
0.60	0.287
0.65	0.340
0.70	0.403
0.75	0.477
0.80	0.567
0.85	0.684
0.90	0.848
0.95	1.129
0.99	1.781

Given:

$$c_v = 0.26 \frac{ft^2}{day} \qquad H_d = 30 ft$$

Note: Drainage height equals full height of clay layer since there is only drainage through the upper gravel layer (one-way drainage).

$U = 0.40 \qquad T_v = 0.126 \qquad$ (from table, $U = 0.40$)

Solve for time: $T_v = \dfrac{c_v \cdot t}{H_d^2}$

Rewriting and solving for t: $\quad t = \dfrac{T_v \cdot H_d^2}{c_v} = \dfrac{0.126 \cdot (30\,ft)^2}{0.26 \dfrac{ft^2}{day}} = 436.15\,days$

Answer: 436

Copyright 2018 by CivilPEPractice
For more problems visit https://civilpepractice.com/

AM PRACTICE EXAM – Version A ANSWERS

4. The maximum shear force (kips) for the beam shown below is most nearly:

   (A) 9.0
   (B) 15.0
   (C) 18.0
   (D) 25.0

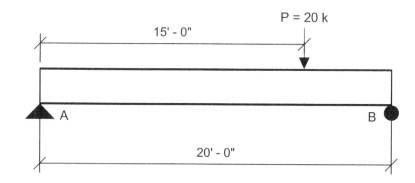

Calculate end reactions:

$\sum$Moments about point B = 0

$(R_A \cdot 20\,ft) - (20k \cdot 5\,ft) = 0$

$R_A = \dfrac{20k \cdot 5\,ft}{20\,ft} = 5k$

$\sum$Forces in y-direction = 0

$R_A + R_B - 20k = 0$

$R_B = 20k - 5k = 15k$

Sketch shear diagram →

Answer: 15.0

5. Fatigue of structural steel is best defined as:

   (A) Deflection exceeding recommended limits
   (B) Excessive shear deformation
   (C) Weakening caused by repeatedly applied loads
   (D) Failure due to excessive bending stress

Fatigue is defined as the weakening of a material due to repeatedly applied loads.

Answer: Weakening caused by repeatedly applied loads

6. The effective vertical stress (psf) at Point A of the soil profile shown is most nearly:

   (A) 2265
   (B) 3265
   (C) 3555
   (D) 3825

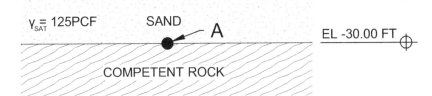

$\gamma_{gravel} = 125 \dfrac{lb}{ft^3}$

$\gamma_{sat1} = 130 \dfrac{lb}{ft^3}$

$\gamma_{sat2} = 125 \dfrac{lb}{ft^3}$ $H_1 = 5 ft$ $H_2 = 15 ft$

$\gamma_{water} = 62.4 \dfrac{lb}{ft^3}$ $H_3 = 10 ft$ $H_{water} = 25 ft$

Calculate effective stress at point A:

$$\sigma'_v = \sum (\gamma_i \cdot H_i) - (\gamma_{water} \cdot H_{water})$$

$$\sigma'_v = \left(125 \dfrac{lb}{ft^3} \cdot 5 ft\right) + \left(130 \dfrac{lb}{ft^3} \cdot 15 ft\right) + \left(125 \dfrac{lb}{ft^3} \cdot 10 ft\right) - \left(62.4 \dfrac{lb}{ft^3} \cdot 25 ft\right)$$

$$\sigma'_v = 2265 \dfrac{lb}{ft^2}$$

Answer: 2265

AM PRACTICE EXAM – Version A ANSWERS

7. A family purchases a vacation home adjacent to a large lake. The family plans on using the vacation home every summer for the next 25 years. The probability that a 100-year flood does not occur over the next 25 years is most nearly:

   (A) 0.222
   (B) 0.546
   (C) 0.778
   (D) 0.990

Given:

$F = 100$

$n = 25$

Calculate probability:

$$P = \left(1 - \frac{1}{F}\right)^n = \left(1 - \frac{1}{100}\right)^{25} = 0.778$$

Answer: 0.778

AM PRACTICE EXAM – Version A ANSWERS

8. A vertical excavation is required at the construction site of a new office building. An 8 ft deep vertical cut is made in deep clay soils. The clay has a unit weight and an unconfined compressive strength of 120 pcf and 2000 psf respectively. The factor of safety of the vertical cut is most nearly:

   (A) 2.08
   (B) 2.25
   (C) 2.35
   (D) 2.45

Given:

$$\gamma = 120 \frac{lb}{ft^3} \qquad q_u = 2000 \frac{lb}{ft^2}$$

Calculate cohesion of clay soil:

$$c = \frac{q_u}{2} = 1000 \frac{lb}{ft^2}$$

Calculate max allowable vertical cut:

$K_a = 1.0$ (for clay soils)

$$z_{cr} = \frac{2 \cdot c}{\gamma \cdot \sqrt{K_a}} = \frac{2 \cdot \left(1000 \frac{lb}{ft^2}\right)}{\left(120 \frac{lb}{ft^3}\right) \cdot \sqrt{1.0}} = 16.67 \, ft$$

Solve for factor of safety:

$$FS = \frac{z_{cr}}{z_{actual}} = \frac{16.67 \, ft}{8.0 \, ft} = 2.08$$

Answer: 2.08

AM PRACTICE EXAM – Version A ANSWERS

9. Speed data was collected as part of a transportation study of a rural town. The data was collected on a moderately traveled two lane collector road. The speed data is given below. The time mean speed (mph) is most nearly:

   (A) 50.00
   (B) 50.15
   (C) 50.23
   (D) 50.36

Recorded Speed (mph)	Frequency
46	16
47	30
48	67
49	105
50	155
51	112
52	84
53	46
54	18

$$TMS = \frac{\sum speed \cdot freq}{\sum freq}$$

Recorded Speed (mph)	Frequency	speed·freq (mph)
46	16	736
47	30	1410
48	67	3216
49	105	5145
50	155	7750
51	112	5712
52	84	4368
53	46	2438
54	18	972
SUM	633	31747

$$TMS = \frac{31747 \, mph}{633} = 50.153 \, mph$$

Answer: 50.15

Copyright 2018 by CivilPEPractice
For more problems visit https://civilpepractice.com/

AM PRACTICE EXAM – Version A ANSWERS

10. The depth of flowing water for the given channel is 3.5 ft. The velocity (fps) of the water is most nearly:

    Given:
    Manning's roughness = 0.025
    Channel slope = 1.0%

    (A) 9.1
    (B) 9.5
    (C) 10.4
    (D) 13.9

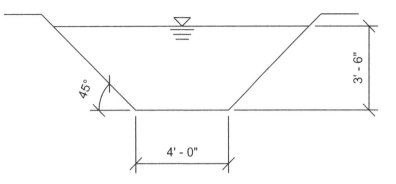

Calculate area of channel:

$$A = \frac{4\,ft + \left[4\,ft + 2 \cdot \left(\frac{3.5\,ft}{\tan 45}\right)\right]}{2} \cdot 3.5\,ft = 26.25\,ft^2$$

Calculate the wetted perimeter:

$$WP = 4\,ft + 2 \cdot \left(\frac{3.5\,ft}{\sin 45}\right) = 13.899\,ft$$

Calculate hydraulic radius:

$$R = \frac{A}{WP} = \frac{26.25\,ft^2}{13.899\,ft} = 1.899\,ft$$

Solve for water velocity using Manning's Equation:

$$V = \frac{1.49}{n} \cdot R^{\frac{2}{3}} \cdot S^{\frac{1}{2}} = \frac{1.49}{0.025} \cdot (1.889\,ft)^{\frac{2}{3}} \cdot (0.01)^{\frac{1}{2}} = 9.11\,fps$$

Note: Slope is expressed as a decimal and all lengths are in feet.

Answer: 9.1

AM PRACTICE EXAM – Version A ANSWERS

11. The ground elevation at point C is most nearly:

   (A) 104.30
   (B) 105.55
   (C) 106.12
   (D) 106.79

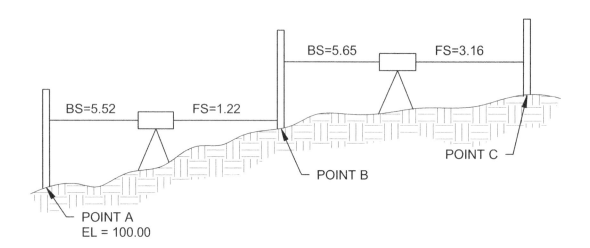

Calculate ground elevation at point B:

$$EL_B = EL_A + BS_A - FS_B = 100.00 + 5.52 - 1.22 = 104.30$$

Solve for ground elevation at point C:

$$EL_C = EL_B + BS_B - FS_C = 104.30 + 5.65 - 3.16 = 106.79$$

Answer: 106.79

AM PRACTICE EXAM – Version A **ANSWERS**

12. A 12-ft thick soil layer has the following properties:

$C_c = 0.14$ $C_r = 0.02$
$e_o = 0.92$ $\sigma'_o = 1,700$ psf
$\sigma'_c = 2,300$ psf

An additional 2,000 psf vertical load will be applied to the top of the soil layer. The primary consolidation (in) is most nearly:

(A) 0.20
(B) 0.50
(C) 1.00
(D) 2.40

Given:

$C_c = 0.14$ $C_r = 0.02$

$e_o = 0.92$ $\sigma'_o = 1,700\,psf$

$\sigma'_c = 2,300\,psf$ $\Delta\sigma = 2,000\,psf$ (additional load)

$H = 12\,ft$

Determine consolidation status:

$\sigma'_c > \sigma'_o \therefore$ soil is overconsolidated

$\sigma'_f = \sigma'_o + \Delta\sigma = 1,700\,psf + 2,000\,psf = 3,700\,psf$

$\sigma'_f > \sigma'_c \therefore \quad S_c = \left(\dfrac{C_r}{1+e_o}\right) \cdot H \cdot \log\left(\dfrac{\sigma'_c}{\sigma'_o}\right) + \left(\dfrac{C_c}{1+e_o}\right) \cdot H \cdot \log\left(\dfrac{\sigma'_f}{\sigma'_c}\right)$

Calculate consolidation:

$S_c = \left(\dfrac{0.02}{1+0.92}\right) \cdot (12\,ft) \cdot \log\left(\dfrac{2,300\,psf}{1,700\,psf}\right) + \left(\dfrac{0.14}{1+0.92}\right) \cdot (12\,ft) \cdot \log\left(\dfrac{3,700\,psf}{2,300\,psf}\right)$

$S_c = 0.197\,ft$ or 2.40in

Answer: 2.40

Copyright 2018 by CivilPEPractice
For more problems visit https://civilpepractice.com/

AM PRACTICE EXAM – Version A ANSWERS

13. The properties for water flowing through a 6-inch (I.D.) closed conduit are given below. The flow is best classified as:

    (A) Laminar
    (B) Transitional
    (C) Turbulent
    (D) None of the above

    Given:

    $v = 0.5 \dfrac{ft}{sec}$

    $\mu = 1 \cdot 10^{-5} \dfrac{ft^2}{sec}$

    $d = 6\,in$

Solve for Reynold's Number:

$$RN = \dfrac{d \cdot v}{\mu} = \dfrac{6in \cdot \left(\dfrac{ft}{12in}\right) \cdot \left(0.5 \dfrac{ft}{sec}\right)}{1 \cdot 10^{-5} \dfrac{ft^2}{sec}} = 25{,}000$$

Note:
RN < 2100, flow is laminar
2100 < RN < 4000, flow is transitional
RN > 4000, flow is turbulent

RN > 4000, therefore flow is turbulent

Answer: Turbulent

AM PRACTICE EXAM – Version A ANSWERS

14. The floor framing shown below supports the following loads:

   6 in normal weight (150 pcf) reinforced concrete slab
   suspended ceiling = 3 psf
   mechanical piping = 5 psf

   The unfactored superimposed dead load (k/ft) on member A is most nearly:

   (A) 0.52
   (B) 0.65
   (C) 0.72
   (D) 1.20

Given:

t_{conc} = 6in

$DL_{ceiling} = \dfrac{3lb}{ft^2}$

$DL_{piping} = \dfrac{5lb}{ft^2}$

Determine the tributary width and calculate the total dead load:

$A_{trib} = \dfrac{25ft}{4\,spaces} = 6.25\,ft$

$w_{conc} = \gamma_{conc} \cdot t_{conc} \cdot A_{trib} = \left(\dfrac{150lb}{ft^3}\right) \cdot \left(\dfrac{6in}{\frac{12in}{ft}}\right) \cdot (6.25ft) = \dfrac{468.75lb}{ft}$

$w_{ceiling} = \dfrac{3lb}{ft^2} \cdot A_{trib} = \left(\dfrac{3lb}{ft^2}\right) \cdot (6.25ft) = \dfrac{18.75lb}{ft}$

$w_{piping} = \dfrac{5lb}{ft^2} \cdot A_{trib} = \left(\dfrac{5lb}{ft^2}\right) \cdot (6.25ft) = \dfrac{31.25lb}{ft}$

$w_{DL-tot} = w_{conc} + w_{ceiling} + w_{piping} = \dfrac{468.75lb}{ft} + \dfrac{18.75lb}{ft} + \dfrac{31.25lb}{ft} = \dfrac{518.75lb}{ft}$ or 0.52 k/ft

Answer: 0.52

AM PRACTICE EXAM – Version A ANSWERS

15. Which of the following work-related incidents must be reported to the Occupational Safety and Health Administration (OSHA)?

    (A) Fatality
    (B) Fatality and loss of eye
    (C) Fatality, amputations, and loss of eye
    (D) Fatality, broken arm resulting in hospitalization, amputations, and loss of eye

    According to OSHA regulation 1904.39 the following are required to be reported:

    - Fatalities
    - Hospitalizations
    - Amputations
    - Loss of eye

    ## Answer: D

16. A steel pipe is reduced from 1' – 6" diameter to 6" diameter as shown below. The flow rate through pipe #1 is 5 cfs. The velocity (ft/sec) through pipe #3 is most nearly:

    (A) 21.1
    (B) 21.6
    (C) 22.4
    (D) 25.5

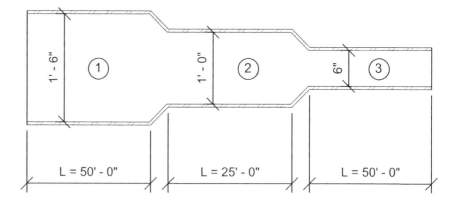

Given:

$Q_1 = 5 \dfrac{ft^3}{\sec}$ $d_1 = 18 in$

$d_2 = 12 in$ $d_3 = 6 in$

Solve for the velocity in pipe #3:

$$A_3 = \frac{\pi \cdot d^2}{4} = \frac{\pi \cdot (6in)^2}{4} = 28.274 \; in^2 \; or \; 0.1963 \; ft^2$$

Continuity equation: $Q_{in} = Q_{out} \therefore Q_3 = Q_1$

$$v_3 = \frac{Q_1}{A_3} = \frac{5 \dfrac{ft^3}{\sec}}{0.1963 \; ft^2} = 25.5 \; \dfrac{ft}{\sec}$$

Answer: 25.5

AM PRACTICE EXAM – Version A ANSWERS

17. A concrete pipe encasement is constructed around a 42 in diameter (O.D.) steel pipe. The steel pipe connects two manholes which are 125 ft apart. The volume of concrete (yd³) required to encase the pipe is most nearly:

(A) 15.4
(B) 71.2
(C) 115.7
(D) 1922.4

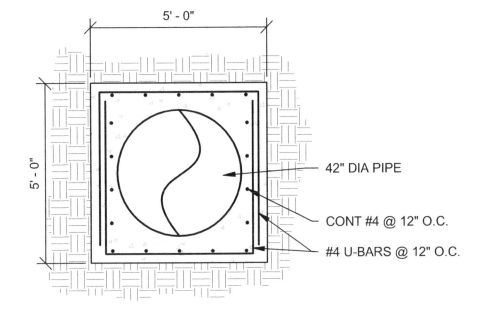

Given:

$d = 42in$

$L = 125ft$

Calculate area of pipe:

$$A_{pipe} = \frac{\pi \cdot d^2}{4} = \frac{\pi \cdot (42in)^2}{4} = 1385.442 in^2 = 9.621 ft^2$$

Solve for the volume of concrete:

$$V_{conc} = \left[(5ft \cdot 5ft) - A_{pipe}\right] \cdot L_{pipe} = \left[(5ft \cdot 5ft) - 9.621 ft^2\right] \cdot 125 ft$$

$$V_{conc} = 1922.4 ft^3 = 71.2 yd^3$$

Answer: 71.2

AM PRACTICE EXAM – Version A ANSWERS

18. A 0.50 ft³ soil sample weighs 65.2 lb. The sample has a moisture content of 0.17 and specific gravity of 2.75. The degree of saturation is most nearly:

   (A) 0.54
   (B) 0.65
   (C) 0.75
   (D) 0.87

Given:

$W = 65.2 \, lb \qquad w = 0.17 \qquad V = 0.50 \, ft^3$

$G_s = 2.75 \qquad \gamma_w = 62.4 \, \dfrac{lb}{ft^3}$

Calculating unit weights:

$\gamma = \dfrac{W}{V} = \dfrac{65.2 \, lb}{0.50 \, ft^3} = 130.4 \, \dfrac{lb}{ft^3}$

$\gamma_d = \dfrac{\gamma}{(1+w)} = 111.453 \, \dfrac{lb}{ft^3}$

Calculating void ratio:

$e = \dfrac{G_s \cdot \gamma_w}{\gamma_d} - 1 = \dfrac{2.75 \cdot \left(62.4 \, \dfrac{lb}{ft^3}\right)}{111.453 \, \dfrac{lb}{ft^3}} - 1 = 0.540$

Solving for degree of saturation:

$S = \dfrac{w \cdot G_s}{e}$

$S = \dfrac{0.17 \cdot 2.75}{0.540} = 0.87$

Answer: 0.87

AM PRACTICE EXAM – Version A ANSWERS

19. A six-story apartment complex is being construction adjacent to a large residential subdivision. Soil borings show 15ft of poor soils over a thick, competent rock layer. Which of the following foundation systems is best suited for the application?

   (A) Spread footings
   (B) Driven piles
   (C) Mat foundation
   (D) Drilled piles

   - Spread footings – Not ideal. A shallow foundation system is inadequate due to poor upper soils.
   - Driven piles – Not ideal. The vibrations created from driving the piles is not desired near a residential subdivision.
   - Mat foundation – Not ideal. A shallow foundation system is inadequate due to poor upper soils.
   - Drilled piles – Ideal. A deep foundation system with minimal vibrations during construction.

Answer: Drilled piles

20. The total float (days) for activity G is most nearly:
    Note: all duration values in the table below are given as days.

    (A) 0
    (B) 1
    (C) 2
    (D) 3

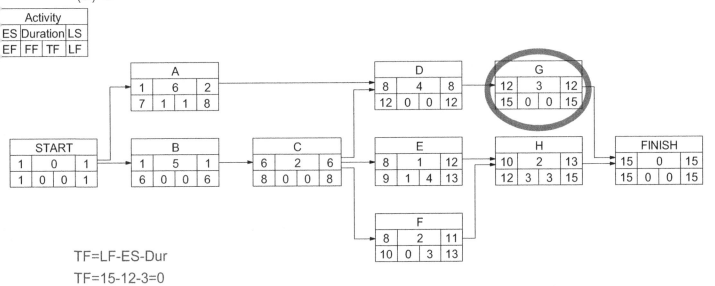

TF=LF-ES-Dur

TF=15-12-3=0

Also, activity G is on the critical path, therefore total float is 0 days.

Answer: 0

AM PRACTICE EXAM – Version A ANSWERS

21. A contractor is preparing to construct a 10,000 square foot steel framed building. Prior to pouring the slab-on-grade the contractor plans to compact the subgrade which is comprised of plastic clays. Which of the following compactors is best suited for the site?

    (A) Sheepsfoot Roller
    (B) Vibratory Plate Compactor
    (C) Vibrating Smooth Wheel Roller
    (D) Pneumatic Tire Roller

    - Sheepsfoot Rollers are best for compacting fine-grained soils such as clays.
    - Vibratory plate compactors are ideal for coarse-grained soils.
    - Vibrating smooth wheel rollers are ideal for upper soil layers and coarse-grained soils such as dense graded aggregate.
    - Pneumatic tire rollers are best used on coarse-grained soils with fines or asphalt pavements.

    ## Answer: Sheepsfoot Rollers

AM PRACTICE EXAM – Version A ANSWERS

22. A 12,000 lb beam is lifted with the rig shown below. The tension force (lb) in Sling B is most nearly:
    Note: assume weight of the beam is uniformly distributed.

    (A) 6000
    (B) 6200
    (C) 6700
    (D) 6800

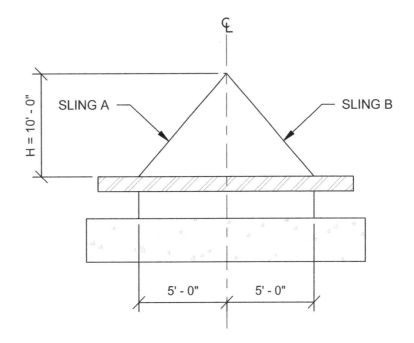

Calculate vertical force acting at sling B:

Since the concrete beam is supported symmetrically, each sling receives one half of the total load

$$F_{y(B)} = \frac{12,000 lb}{2} = 6,000 lb$$

Calculate tension force in sling B:

$$T_B = \frac{\sqrt{(5 ft)^2 + (10 ft)^2}}{10 ft} \cdot 6,000 lb = 6,708 lb$$

Answer: 6,700

AM PRACTICE EXAM – Version A ANSWERS

23. Which of the following is utilized as a sediment control method?

    (A) Geotextiles
    (B) Straw mulch
    (C) Silt fence
    (D) Hydraulic mulch

Sediment control methods include <u>silt fences</u>, sediment basins, sediment traps and similar mechanisms to trap soil particles once the soil has eroded.

The other listed answers are methods used to prevent erosion altogether which are called erosion controls.

Answer: Silt fence

24. The contractor selects an excavator with a bucket capacity of 4-yd³ and a standard operating cycle time of 50 seconds for his project. The number of excavators required to excavate 7500 yd³ of soil in 8 hours is most nearly:

    (A) 3
    (B) 4
    (C) 5
    (D) 6

Given:

$t = 8hr$ $\quad$ $t_{cycle} = 50 sec$ $\quad$ $v_{cycle} = 4 yd^3$ $\quad$ $v_{tot} = 7{,}500 yd^3$

Calculate the 8 hour production of one excavator:

$$\text{Prod} = \frac{v_{cycle} \cdot t}{t_{cycle}} = \frac{(4yd^3)\cdot(8hr)\cdot\left(\frac{3600\sec}{hr}\right)}{50\sec} = 2{,}304\,yd^3$$

Solve number of excavators required:

$$excavators = \frac{v_{tot}}{\text{Prod}} = \frac{7{,}500\,yd^3}{2{,}304\,yd^3} = 3.255$$

Answer: 4 excavators

AM PRACTICE EXAM – Version A ANSWERS

25. The maximum bearing pressure (psf) for the footing shown below is most nearly:

Note: all loads are unfactored.

(A) 5775
(B) 5850
(C) 5925
(D) 6250

Given:
B = 3.5 ft
L = 4 ft
P = 40,000 lb
M = 25,000 lb-ft

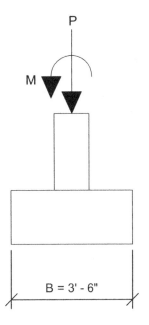

Calculate eccentricity due to moment acting on footing:

$$e = \frac{M}{P} = \frac{25,000 \, lb \cdot ft}{40,000 \, lb} = 0.625 \, ft$$

Solve for maximum bearing pressure:

$$Q_{max} = \frac{P}{B \cdot L} \cdot \left(1 + \frac{6 \cdot e}{B}\right) \text{ for } e < \frac{B}{6}$$

$$Q_{max} = \frac{4 \cdot P}{3 \cdot L \cdot (B - 2e)} \text{ for } e > \frac{B}{6}$$

$$\frac{B}{6} = \frac{3.5 \, ft}{6} = 0.583 \, ft$$

$$e > \frac{B}{6} \therefore Q_{max} = \frac{4 \cdot P}{3 \cdot L \cdot (B - 2e)}$$

$$Q_{max} = \frac{4 \cdot P}{3 \cdot L \cdot (B - 2e)} = \frac{4 \cdot (40,000 \, lb)}{3 \cdot (4 \, ft) \cdot (3.5 \, ft - 2(0.625 \, ft))} = 5925.93 \frac{lb}{ft^2}$$

Answer: 5925

Copyright 2018 by CivilPEPractice

AM PRACTICE EXAM – Version A ANSWERS

26. Which of the following diagrams best matches the activity data found in the table below?

   (A) Diagram A
   (B) Diagram B
   (C) Diagram C
   (D) Diagram D

Activity	Successor	Duration (days)
Start (A)	B,C	
B	D	6
C	D,E	8
D	F	4
E	F	3
F	G,H	4
G	J	6
H	J	7
Finish (J)		

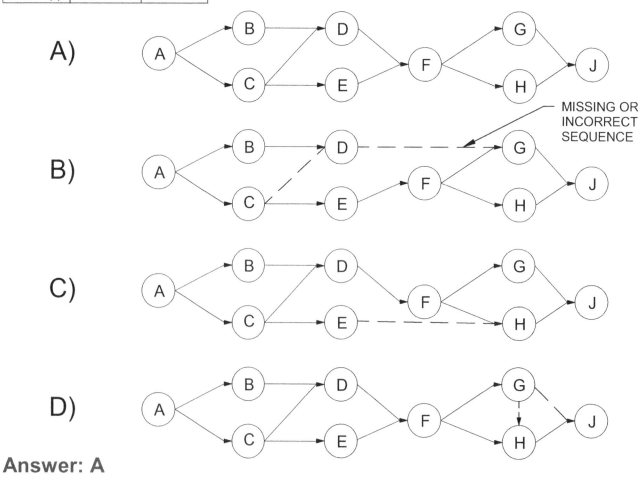

Answer: A

27. The slenderness ratio for the steel column shown below is most nearly:

(A) 8.16
(B) 48.98
(C) 73.47
(D) 97.96

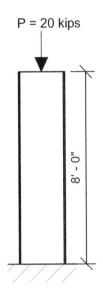

Column Properties	
Weight	50 plf
Area	14.7 in²
Depth	12 in
Radius of gyration	1.96 in

Calculate slenderness ratio:

$$SR = \frac{k \cdot L}{r}$$

k = 2.0 for fixed column base and free top (CERM 45 – 3)

L = 8ft

r = 1.96in

$$SR = \frac{k \cdot L}{r} = \frac{2.0 \cdot (8\,ft)}{1.96\,in \cdot \left(\frac{ft}{12\,in}\right)} = 97.96$$

Answer: 97.96

AM PRACTICE EXAM – Version A ANSWERS

28. The length of tangent (feet) for a horizontal simple circular curve with a radius equal to 2500 feet and an intersection angle of 80 degrees is most nearly:

    (A) 1000
    (B) 1500
    (C) 2100
    (D) 2800

Given:

$R = 2500\,ft$

$I = 80$

Solve for the length of tangent

$$T = R \cdot \tan\frac{I}{2} = 2500 ft \cdot \tan\left(\frac{80}{2}\right) = 2097.75\,ft$$

Answer: 2100

AM PRACTICE EXAM – Version A ANSWERS

29. A continuous 3 ft wide footing bears on a thick layer of clean sand. The bottom of footing is located 4 ft below grade. Use Terzaghi's method to determine the maximum net allowable bearing capacity (psf):

   (A) 3500
   (B) 4500
   (C) 5500
   (D) 6500

Given:

$$\gamma = 125 \frac{lb}{ft^3} \qquad \phi = 30 \qquad FS = 3.0$$

$$N_c = 37.16 \qquad N_q = 22.46 \qquad N_\gamma = 19.73$$

Determine shape factors for footing (See Civil Engineering Reference Manual Chapter 36):

$$S_\gamma = 1.0 \ \& \ S_c = 1.0$$

Solve for gross ultimate bearing capacity:

$$q_{ult-gross} = (c \cdot N_c \cdot S_c) + (\gamma \cdot D_f \cdot N_q) + (0.5 \cdot \gamma \cdot B \cdot N_\gamma \cdot S_\gamma)$$

$D_f = 4\ ft$ (depth of footing) $B = 3\ ft$ (width of footing)

$c = 0$ (cohesion of a *clean sand*)

$$q_{ult-gross} = (0 \cdot 37.16 \cdot 1.0) + (125 \frac{lb}{ft^3} \cdot 4\ ft \cdot 22.46) + (0.5 \cdot 125 \frac{lb}{ft^3} \cdot 3\ ft \cdot 19.73 \cdot 1.0)$$

$$q_{ult-gross} = 14{,}929 \frac{lb}{ft^2}$$

Solve for net ultimate bearing capacity:

$$q_{ult-net} = q_{ult-gross} - (\gamma \cdot D_f) = 14{,}929 \frac{lb}{ft^2} - (125 \frac{lb}{ft^3} \cdot 4\ ft) = 14{,}429 \frac{lb}{ft^2}$$

Solve for net allowable bearing capacity.

$$q_{allow-net} = \frac{q_{ult-net}}{FS} = \frac{14{,}429 \frac{lb}{ft^2}}{3.0} = 4{,}810 \frac{lb}{ft^2}$$

Answer: 4,500

AM PRACTICE EXAM – Version A ANSWERS

30. The force (kips) in member CH of the truss shown below is most nearly:

   (A) 18.4 Tension
   (B) 18.4 Compression
   (C) 20.0 Tension
   (D) 20.0 Compression

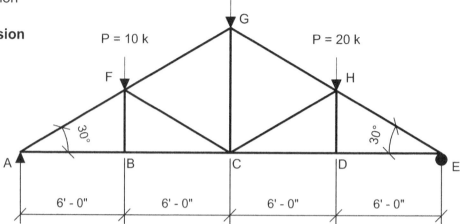

Solve for reactions at point E:

$$\sum M_A = 0$$

$$(10\,k \cdot 6\,ft) + (30\,k \cdot 12\,ft) + (20\,k \cdot 18\,ft) - (E_y \cdot 24\,ft) = 0$$

$$E_y = 32.5k$$

$$E_x = 0k \text{ (roller support)}$$

Using method of sections solve for force in member CH:

Sum Moments about point E:

$$F_{CHy} \cdot (12\,ft) - 20\,k \cdot (6\,ft) = 0$$

$$F_{CHy} = \frac{20\,k \cdot 6\,ft}{12\,ft} = 10.0k$$

$$F_{CH} = \frac{F_{CHy}}{\sin 30} = 20.0k$$

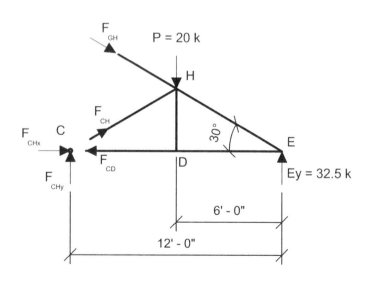

Answer: 20.0 Compression

AM PRACTICE EXAM – Version A ANSWERS

31. A cylindrical tank has a diameter of 18ft and is 16ft deep. The flow rate through the tank is 200 gpm. The detention time (hrs) is most nearly:

    (A) 2.54
    (B) 2.65
    (C) 116.52
    (D) 152.28

Given:

$d = 18ft$ $\qquad$ $h = 16ft$ $\qquad$ $Q = 200 \dfrac{gal}{min}$

Calculate tank volume:

$$V = \frac{\pi \cdot d^2}{4} \cdot h = \frac{\pi \cdot (18\,ft)^2}{4} \cdot 16\,ft = 4071.5\,ft^3$$

Solve for detention time:

$$t_d = \frac{V}{Q} = \frac{4071.5\,ft^3}{\left(200\dfrac{gal}{min}\right) \cdot \left(\dfrac{ft^3}{7.48\,gal}\right)} = 152.27\,min = 2.54\,hrs$$

Answer: 2.54

32. The water/cement ratio for the following concrete mix is most nearly:

    (A) 0.53
    (B) 0.50
    (C) 0.45
    (D) 0.40

Material	Material Type	ASTM	Weight (lbs)
Cement	Type I-II	C 150	490
Fly Ash	Fly Ash F	C - 618	95
Coarse Aggregate	# 57	C 33	1724
Fine Aggregate	Natural Sand	C 33	1285
Water	City		262
Admixture	Air Entrainer	C 260	3.5 oz/CY
Admixture	Type A Water Reducer	C 260	4.0 oz/CY

Calculate w/c ratio

$$w/c = \frac{w_{water}}{w_{cement} + w_{flyash}} = \frac{262\,lb}{490\,lb + 95\,lb} = 0.45$$

Answer: 0.45

Copyright 2018 by CivilPEPractice
For more problems visit https://civilpepractice.com/

AM PRACTICE EXAM – Version A ANSWERS

33. The total material cost of concrete for the tank shown below is most nearly:

   Note: Assume a concrete material unit cost of $200 per cubic yard.

   (A) $25,600
   (B) $21,400
   (C) $20,400
   (D) $18,400

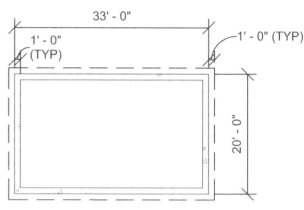

PLAN

Calculate volume of concrete walls:

$$V_{walls} = [(33\,ft \cdot 20\,ft) - (31\,ft \cdot 18\,ft)] \cdot 17\,ft^3$$

$$V_{walls} = 1734\,ft^3 = 64.22\,yd^3$$

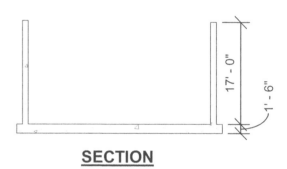

SECTION

Calculate volume of concrete mat foundation:

$$V_{fdn} = (35\,ft \cdot 22\,ft) \cdot 1.5\,ft = 1155\,ft^3$$

or $42.78\,yd^3$

Total volume:

$$V_{tot} = V_{walls} + V_{fdn} = 64.22\,yd^3 + 42.78\,yd^3 = 107\,yd^3$$

Total Cost: $\quad Cost_{tot} = 107\,yd^3 \cdot \left(\dfrac{\$200}{yd^3}\right) = \$21,400$

Answer: $21,400

AM PRACTICE EXAM – Version A ANSWERS

34. A 500' long trench is to be cut to install steel piping. The area to be excavated is shown below. The total volume (yd³) of the excavation is most nearly:

   (A) 1900
   (B) 2000
   (C) 2100
   (D) 2200

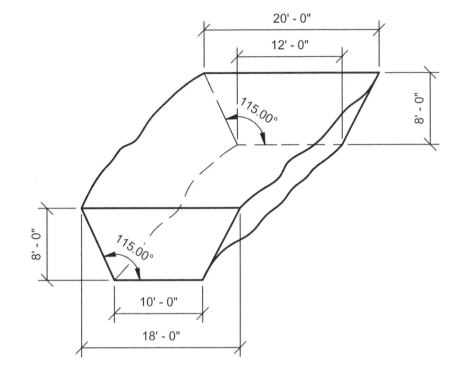

Solve volume of excavation using end area method:

$$V = \frac{L \cdot (A_1 + A_2)}{2}$$

$$A_1 = \frac{(18\,ft + 10\,ft)}{2} \cdot 8\,ft = 112\,ft^2$$

$$A_2 = \frac{(20\,ft + 12\,ft)}{2} \cdot 8\,ft = 128\,ft^2$$

$$L = 500\,ft$$

$$V = \frac{500\,ft \cdot (112\,ft^2 + 128\,ft^2)}{2} = 60{,}000\,ft^3 \text{ or } 2222.22\,yd^3$$

Answer: 2200

AM PRACTICE EXAM – Version A ANSWERS

35. A concrete gravity wall with flat backfill retains 12' of sand. The active resultant force (lb/ft) induced on the wall due to lateral earth pressure is most nearly:

   (A) 374
   (B) 775
   (C) 2244
   (D) 2525

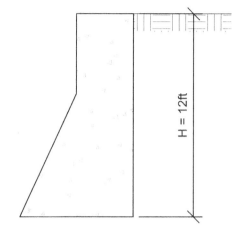

Soil Properties:

$\gamma_{soil} = 115 \dfrac{lb}{ft^3}$

$\phi = 35$

Given:

$\gamma_{soil} = 115 \dfrac{lb}{ft^3}$

$\phi = 35$

H = 12ft

Calculate active pressure coefficient:

$K_a = \tan\left(45 - \dfrac{\phi}{2}\right)^2 = 0.271$

Calculate active pressure:

$p_a = \gamma \cdot H \cdot K_a = 115 \dfrac{lb}{ft^3} \cdot 12\, ft \cdot 0.271 = 373.98 \dfrac{lb}{ft^2}$

Calculate active resultant force:

$R_a = \dfrac{1}{2} \cdot p_a \cdot H = \dfrac{1}{2} \cdot 373.98 \dfrac{lb}{ft^2} \cdot 12\, ft = 2243.88 \dfrac{lb}{ft}$

Answer: 2244

AM PRACTICE EXAM – Version A ANSWERS

36. The contractor prepares the area to perform a concrete slump test. Which of the following is an acceptable surface in which to conduct the test?

    (A) 5/8" solid sheet of plywood
    (B) 4ft x 4ft level gravel surface
    (C) 1/8" thick steel plate
    (D) All of the above

Slump tests shall be conducted on a flat, moist, nonabsorbent (rigid) surface.

Answer: 1/8" thick steel plate

37. The USCS classification of the following soil sample is:

    (A) SC
    (B) SM
    (C) SP
    (D) GC

Sieve Size	Percent Finer
No. 4	90
No. 10	80
No. 40	70
No. 200	40

PL	16
LL	42

Step 1) Percent retained on No. 200 sieve = 60, therefore soil is coarse-grained.

Step 2) Percent retained on No. 4 sieve = 10, therefore soil is sand.

Step 3) More than 12% pass No. 200 sieve, therefore soil is SM, SM-SC, or SC.

Step 4) Plot PI vs LL.

PI = LL – PL = 42 -16 = 26

Point is above "A-Line," therefore soil is SC.

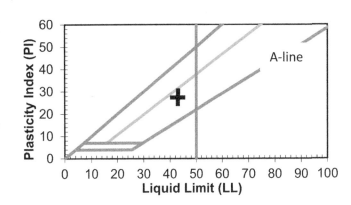

Answer: SC

AM PRACTICE EXAM – Version A ANSWERS

38. A 6-inch diameter (I.D.) pipe is used for a 5,500 ft segment of a water distribution system. The design velocity and friction factor are 6.5 ft/s and 0.0160, respectively. The friction loss (ft) of the pipe segment is most nearly:

    (A) 102
    (B) 107
    (C) 115
    (D) 126

Given:

$$v = 6.5 \frac{ft}{\sec}$$

$$g = 32.2 \frac{ft}{\sec^2}$$

L = 5,500ft

D = 6in

f = 0.0160

Solve friction head loss:

$$h_f = \frac{f \cdot L \cdot v^2}{2 \cdot D \cdot g} = \frac{(0.0160) \cdot (5,500\,ft) \cdot \left(6.5\frac{ft}{\sec}\right)^2}{2 \cdot (6in) \cdot \left(\frac{ft}{12in}\right) \cdot \left(32.2\frac{ft}{\sec^2}\right)} = 115.466\,ft$$

Answer: 115

Copyright 2018 by CivilPEPractice
For more problems visit https://civilpepractice.com/

AM PRACTICE EXAM – Version A ANSWERS

39. The axial force (lb) resisted by a single temporary brace is most nearly:

Note: assume wall base is pinned and the brace only resists lateral load.

(A) 1745
(B) 2468
(C) 2577
(D) 2596

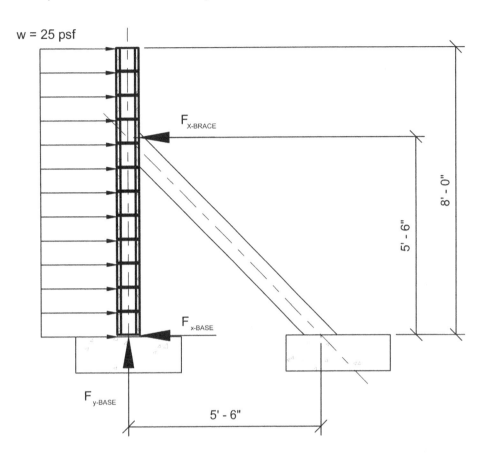

Given:

$w = 25 \dfrac{lb}{ft^2}$

$s = 12\, ft$

$h_{wall} = 8\, ft$

Determine the lateral reaction at the wall/brace interface:

Sum moments @ base of wall = 0

$$\left(F_{x-BRACE} \cdot 5.5\, ft\right) - \left[\left(25\dfrac{lb}{ft^2} \cdot 12\, ft\right) \cdot (8\, ft) \cdot \left(\dfrac{8\, ft}{2}\right)\right] = 0$$

Rewriting and solving:

$$F_{x-BRACE} = \dfrac{\left[\left(25\dfrac{lb}{ft^2} \cdot 12\, ft\right) \cdot (8\, ft) \cdot \left(\dfrac{8\, ft}{2}\right)\right]}{5.5\, ft} = 1745.45\, lb$$

Solve for axial load resisted by brace:

$$F_{axial} = F_{x-BRACE} \cdot \dfrac{\sqrt{(5.5\, ft)^2 + (5.5\, ft)^2}}{5.5\, ft} = 2468.445\, lb$$

Answer: 2468

AM PRACTICE EXAM – Version A ANSWERS

40. A catch basin is to be installed adjacent to a 3.4 acre parking lot. The design rainfall intensity is 6.7 in/hr and the runoff coefficient is 0.84. The peak discharge (ft³/sec) to the catch basin is most nearly:

    (A) 15.4
    (B) 16.3
    (C) 16.7
    (D) 19.1

Given:

C = 0.84

$I = 6.7 \dfrac{in}{hr}$

A = 3.4 acres

Solve for peak discharge:

$Q = C \cdot I \cdot A = 0.84 \cdot \left(6.7 \dfrac{in}{hr}\right) \cdot (3.4 \, acres) = 19.135 \text{ cfs}$

Note the following required units:

C = unitless

I = in/hr

A = acres

Answer: 19.1

Breadth Exam

Version B

AM PRACTICE EXAM – Version B

1. Ceiling fans are to be installed in a new 500 room apartment complex. Each room receives one ceiling fan. The standard productivity for the installing crew is 45-minutes per fan. The contractor uses 12 crews. The duration of the ceiling fan installation (working days) is most nearly:

   Note: assume 8 hr working days.

   (A) 3.2
   (B) 3.6
   (C) 3.9
   (D) 4.2

2. The factor of safety for overturning for the given retaining wall is most nearly:
   Note: ignore the weight of soil over the toe.

   (A) 3.84
   (B) 3.95
   (C) 4.55
   (D) 4.75

   $\gamma_{conc} = 150 \dfrac{lb}{ft^3}$

   $\gamma_{soil} = 115 \dfrac{lb}{ft^3}$

   $K_a = 0.33$

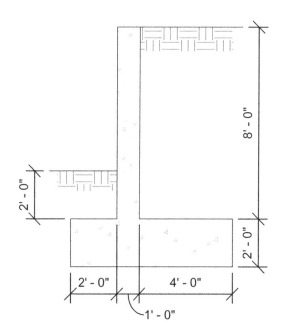

AM PRACTICE EXAM – Version B

3. The maximum unfactored shear stress (psi) for the beam shown below is most nearly:

   (A) 50.00
   (B) 36.67
   (C) 33.33
   (D) 20.00

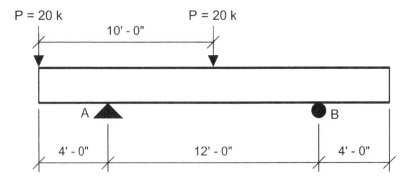
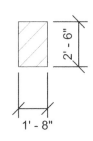

BEAM CROSS SECTION

4. A vertical excavation is to be performed in a deep clay soil. The clay has a unit weight and an unconfined compressive strength of 110 pcf and 2250 psf respectively. The maximum vertical cut (ft) without requiring a lateral support is most nearly:

   (A) 20.45
   (B) 25.20
   (C) 30.67
   (D) 40.91

AM PRACTICE EXAM – Version B

5. A normally consolidated clay soil is subjected to the surcharge shown below. The compression and recompression indexes are 0.10 and 0.05 respectively. The settlement (in) due to the surcharge is most nearly:

   (A) 2.8
   (B) 3.5
   (C) 3.7
   (D) 3.9

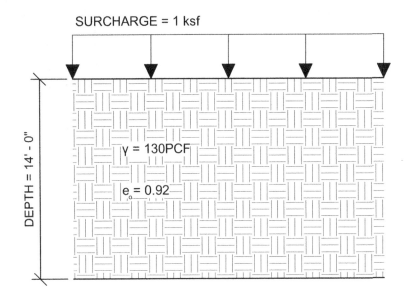

6. A saturated soil specimen has the following properties:

   Weight before oven drying: 3.5 lb
   Weight after oven drying: 3.2 lb
   Specific gravity: 2.5

   The void ratio of the soil specimen is most nearly:

   (A) 0.15
   (B) 0.18
   (C) 0.20
   (D) 0.23

AM PRACTICE EXAM – Version B

7. A falling-head permeability test was performed on a clayey silt. The data collected is shown in the table below. The coefficient of permeability (cm/sec) is most nearly:

   (A) 1.911E-5
   (B) 1.712E-5
   (C) 1.655E-5
   (D) 1.655E-6

Experiment #1	
Area of specimen	85.00 cm²
Length of specimen	15.25 cm
Apparatus Standpipe Area	2.00 cm²
Water Height (start)	100 cm
Water Height (end)	88 cm
Test duration	40 min
Temperature	30°C

8. Which of the following should be added to the concrete mix to best protect against freeze-thaw cycles?

   (A) Retarding admixture
   (B) Accelerating admixture
   (C) Type III cement
   (D) Air-entraining admixture

AM PRACTICE EXAM – Version B

9. A roadway with a roughness coefficient of 0.014 is shown below. The longitudinal slope of the roadway is 2.0%. The minimum width of gutter (ft) required to contain a design flow of 3.0 cfs in its entirety is most nearly:

    (A) 5.0
    (B) 5.5
    (C) 6.5
    (D) 10.0

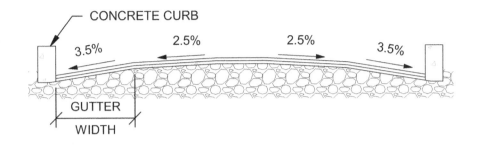

10. Traffic flow was observed along an arterial street. The vehicles were traveling an average of 55mph and were spaced 500 feet apart. The flow of traffic (veh/hr) is most nearly:

    (A) 240
    (B) 560
    (C) 580
    (D) 680

AM PRACTICE EXAM – Version B

11. Which members from the truss below are zero-force members?

    (A) Member CG
    (B) Member CH and Member CF
    (C) Member BF and Member DH
    (D) Member BF, Member DH, Member CG

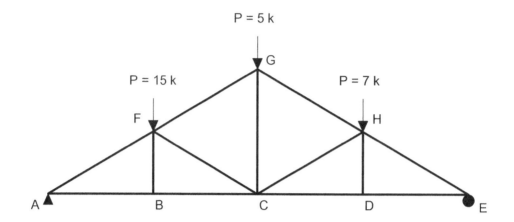

12. A pressure drop of 12 psig is measured across a 150 ft long ductile iron pipe. The water velocity (ft/sec) in the pipe is most nearly:
    Note: the change in elevation and velocity heads are negligible.

    Pipe Properties:
    Inside Diameter = 6 in
    friction coefficient = 140

    (A) 20
    (B) 18
    (C) 17
    (D) 15

AM PRACTICE EXAM – Version B

13. The flow through the given open channel is most nearly characterized by:

    Given:
    V= 10ft/sec

    (A) Critical flow
    (B) Supercritical flow
    (C) Subcritical flow
    (D) None of the above

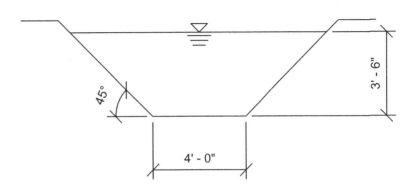

14. A working pressure of 45 psig is needed at Point B of an 8-inch (I.D.) steel pipe. The velocity and friction factor are 6.0 ft/s and 0.0165, respectively. The pressure head (ft) at Point A is most nearly:

    Note: Minor losses are negligible.

    (A) 190
    (B) 175
    (C) 170
    (D) 160

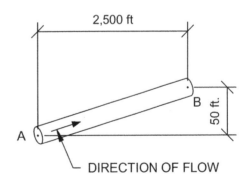

AM PRACTICE EXAM – Version B

15. A -2.75% grade intersects a +1.50% grade at Sta. 20+00 and elevation 200.00 ft. A 300-ft vertical curve connects the two grades. The elevation (feet) of the turning point is most nearly:

    (A) 201.46
    (B) 201.14
    (C) 200.14
    (D) 200.00

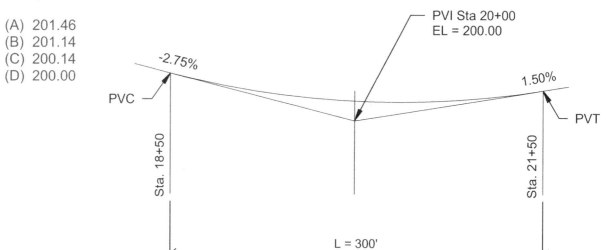

16. A 5.0 acre concrete parking lot is to be constructed. Which of the following compactors is best suited to compact and finish the base layer comprised of dense graded aggregate?

    (A) Sheepsfoot Roller
    (B) Tamping Rollers
    (C) Vibrating Smooth Wheel Roller
    (D) Pneumatic Tire Roller

AM PRACTICE EXAM – Version B

17. The framing for a mechanical mezzanine is shown below. The top of framing is 15'-0" above the foundation. The weight of the mezzanine framing (tons) is most nearly:

    (A) 6.5
    (B) 9.0
    (C) 13.1
    (D) 18.0

    Note: include weight of columns and beams.

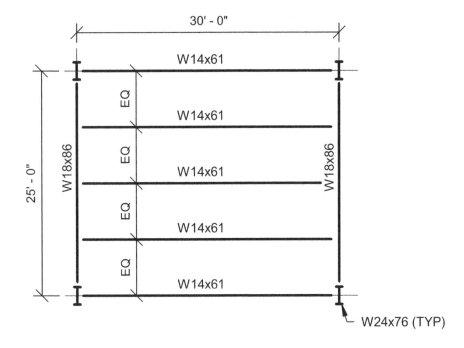

18. An investor purchases a commercial building in a large city adjacent to the main river. The investor takes out a 20-year mortgage to finance the building. The probability of at least one 100-year flood occurring during the mortgage period is most nearly:

    (A) 0.99
    (B) 0.74
    (C) 0.30
    (D) 0.18

19. A circular pipe is to be installed as part of a residential sewer system. The pipe will be installed on a 2% grade and has a roughness coefficient of 0.012. The minimum full-flow capacity shall be 3.2 ft³/sec. The minimum inside diameter of pipe (in) that can meet this requirement is most nearly:

(A) 8
(B) 10
(C) 12
(D) 18

20. The maximum unfactored bending stress (ksi) of the W24×76 beam shown below is most nearly:

Note: Neglect self-weight of beam. All loads given are unfactored.

(A) 12
(B) 16
(C) 20
(D) 25

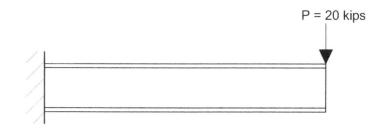

Given:

$S_x = 176 in^3$

$S_y = 18.4 in^3$

Length of beam = 12'-0"

21. The unfactored axial stress (ksi) in column A is most nearly:

   (A) 0.20
   (B) 1.50
   (C) 4.25
   (D) 5.25

Loads:
DL = 40 psf (includes self-weight)
LL = 100 psf

Column properties:
d = 12 in
$b_f = 12$ in
$A = 5\ in^2$

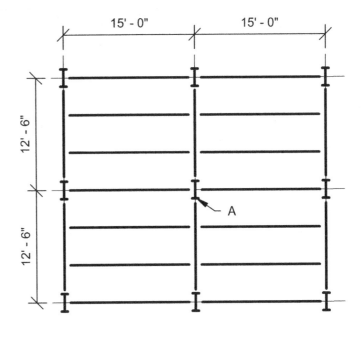

AM PRACTICE EXAM – Version B

22. The length of curve AB (ft) measured along its arc is most nearly:

   (A) 698
   (B) 705
   (C) 812
   (D) 819

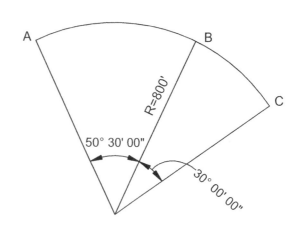

23. Which of the following concrete beams offer the greatest shear resistance to load P?

   (A) A
   (B) B
   (C) C
   (D) D

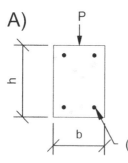
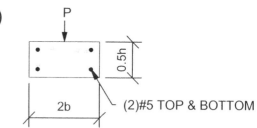
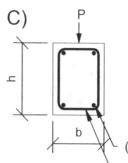
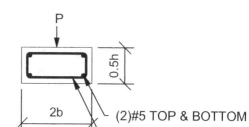

AM PRACTICE EXAM – Version B

24. The survey crew gathers information at a future construction site. A summary of the data collected is given below. The ground elevation at point A is most nearly:

    (A) 93.21
    (B) 94.12
    (C) 95.55
    (D) 97.51

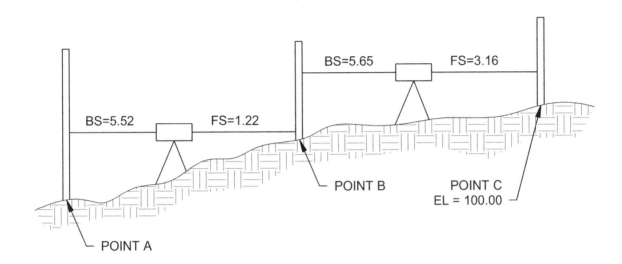

25. Total deflection for the beam shown below shall not exceed L/360 or 0.5 in. What is the lightest beam that will satisfy the deflection requirements?

Note: Neglect self-weight of beam. E = 29,000 ksi

(A) W18x65
(B) W18x71
(C) W18x76
(D) W18X86

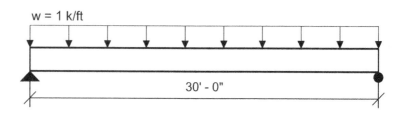

W18x65: $I_x = 1070 in^4$
W18x71: $I_x = 1170 in^4$
W18x76: $I_x = 1330 in^4$
W18X86: $I_x = 1530 in^4$

26. A rectangular wastewater tank has dimensions of 40 ft x 60 ft x 15 ft deep and is receiving a flow of 600 gal/min. The temperature and unit weight of the wastewater are 70°F and 65pcf respectively. The hydraulic detention time (hr) for the tank is most nearly:

(A) 7.12
(B) 7.48
(C) 7.65
(D) 7.76

AM PRACTICE EXAM – Version B

27. A soil profile is given below. The difference in total vertical stresses (psf) at Point A and Point B is most nearly:

(A) 2850
(B) 2784
(C) 2551
(D) 1903

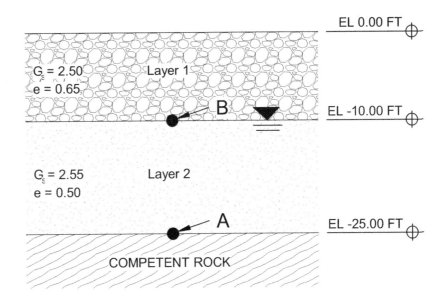

28. A 20-ft thick clay soil layer is confined between two sand layers. The clay has a coefficient of consolidation of 0.28 ft²/day. The time (days) when 50% of the total clay settlement will occur is most nearly:

(A) 25
(B) 70
(C) 160
(D) 281

U=degree of consolidation
T_v=time factor

U	Tv
0.10	0.008
0.20	0.031
0.30	0.071
0.40	0.126
0.50	0.197
0.55	0.238
0.60	0.287
0.65	0.340
0.70	0.403
0.75	0.477
0.80	0.567
0.85	0.684
0.90	0.848
0.95	1.129
0.99	1.781

AM PRACTICE EXAM – Version B

29. A concrete wall retains 16' of sand as shown below. The at-rest resultant (lb/ft) induced on the wall due to lateral earth pressure is most nearly:

    (A) 1,499
    (B) 10,550
    (C) 11,994
    (D) 12,202

$\gamma_{sat} = 125 \dfrac{lb}{ft^3}$

$\varphi' = 30°$

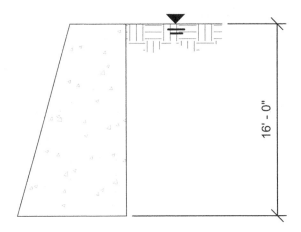

30. According to OSHA regulations which of the following is an acceptable means of fall protection for an employee working on a roof structure 20ft above the ground level?

    (A) guardrail systems
    (B) safety net systems
    (C) fall protection is not required
    (D) guardrail systems and safety net systems

31. The total vertical stress (psf) at Point A of the soil profile shown is most nearly:

(A) 850
(B) 1350
(C) 1750
(D) 2600

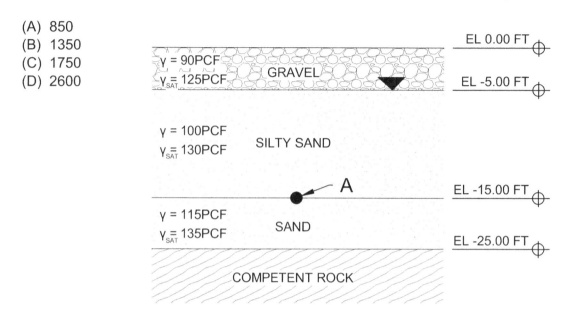

32. Concrete creep is best defined as:

(A) weakening caused by repeatedly applied loads.
(B) contracting of a hardened concrete mixture due to the loss of capillary water.
(C) deterioration of concrete due to freeze thaw cycles.
(D) time-dependent deformation under a sustained load.

AM PRACTICE EXAM – Version B

33. The net excavation (yd³) from Station 500+00 to 507+00 is most nearly:

    (A) 1400 cut
    (B) 2000 cut
    (C) 1400 fill
    (D) 2000 fill

Station	End Area (ft²) Cut	End Area (ft²) Fill
500 + 00	0	325
501 + 00	0	225
502 + 00	0	300
503 + 00	0	0
504 + 00	125	0
505 + 00	450	0
506 + 00	375	0
507 + 00	250	0

34. Concrete forms shall be designed for which of the following loads:

    (A) Weight of concrete and Construction Live Loads
    (B) Hydrostatic pressure of wet concrete
    (C) Both A and B
    (D) None of the Above

AM PRACTICE EXAM – Version B

35. A contractor is required to install 12 bollards outside of a new fire station. It takes 2 laborers 2 hours to install one bollard. The pay rate for a single laborer is $25/hr. The total cost to install all 12 bollards including labor and material is most nearly:

(A) $375
(B) $1200
(C) $3300
(D) $4500

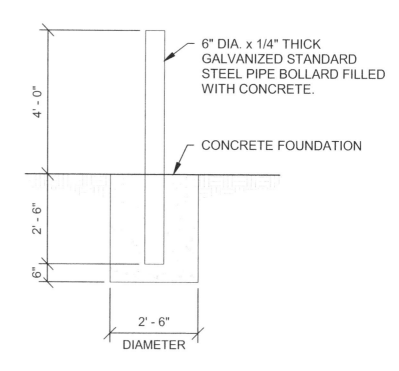

Material	Unit	Unit Cost
Steel Pipe	LF	$25
Concrete Foundation	CY	$200
Concrete Fill Inside Pipe	CY	$150

AM PRACTICE EXAM – Version B

36. A partial plan of the formwork for a suspended 8" thick concrete slab is given below. The girders are spaced at 8'-0" on center and the joists are spaced at 2'-0" on center. The superimposed live load on the formwork is 50psf. The maximum unfactored bending moment (lb-ft) in the formwork joist is most nearly:

Note: ignore self-weight of the joists

(A) 1215
(B) 1615
(C) 2430
(D) 3230

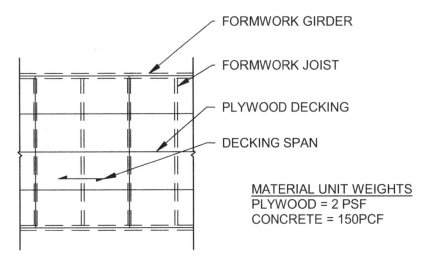

FORMWORK GIRDER
FORMWORK JOIST
PLYWOOD DECKING
DECKING SPAN

MATERIAL UNIT WEIGHTS
PLYWOOD = 2 PSF
CONCRETE = 150PCF

37. A contractor is to construct a building addition for a shipping and receiving center. The foundation for the addition is to be constructed 5 feet above the existing building foundation. The new addition is to match the existing construction with shallow spread footings and concrete stem walls. Which of the following construction methods should be used to prevent overloading the existing footing?

(A) Dowel the new foundation into the existing
(B) Construct the new foundation outside the zone of influence of the existing footing
(C) Both A and B
(D) None of the Above

38. Which of the following is utilized as an erosion control method?

(A) Geotextiles
(B) Straw mulch
(C) Hydraulic mulch
(D) All of the Above

AM PRACTICE EXAM – Version B

39. A critical path diagram is shown below for an industrial construction project. The minimum number of days required to complete the project is most nearly:

(A) 22
(B) 23
(C) 25
(D) 27

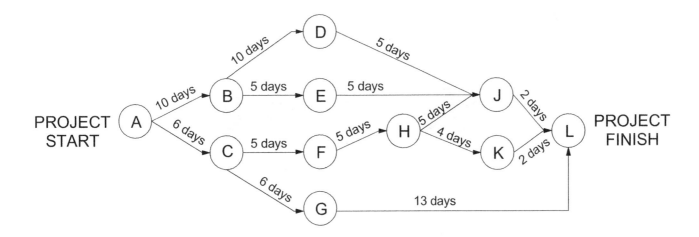

40. An excavator with a bucket capacity of 3-yd³ has a standard operating cycle time of 55 seconds. One 24 yd³ dump truck is used to dispose of the excavated soil generated by the excavator. The total number of trips required to haul all of the excavated soil generated by one excavator during an 8 hour workday is most nearly:

(A) 60
(B) 65
(C) 66
(D) 68

Breadth Exam

Version B

Answers

AM PRACTICE EXAM – Version B ANSWERS

1. Ceiling fans are to be installed in a new 500 room apartment complex. Each room receives one ceiling fan. The standard productivity for the installing crew is 45-minutes per fan. The contractor uses 12 crews. The duration of the ceiling fan installation (working days) is most nearly:

   Note: assume 8 hour working days.

   (A) 3.2
   (B) 3.6
   (C) 3.9
   (D) 4.2

$$\left(\frac{500\ rooms}{project}\right)\cdot\left(\frac{1\ fan}{room}\right)\cdot\left(\frac{\frac{45\ min}{fan}}{crew}\right)\cdot\left(\frac{1\ hr}{60\ min}\right)\cdot\left(\frac{project}{12\ crews}\right)\cdot\left(\frac{1\ day}{8\ hrs}\right)=3.906\ days$$

Answer: 3.9

2. The factor of safety for overturning for the given retaining wall is most nearly:
   Note: ignore the weight of soil over the toe.

   (A) 3.84
   (B) 3.95
   (C) 4.55
   (D) 4.75

$$\gamma_{conc} = 150\,\frac{lb}{ft^3}$$

$$\gamma_{soil} = 115\,\frac{lb}{ft^3}$$

$$K_a = 0.33$$

Wall properties:

$$h = 8\,ft + 2\,ft = 10\,ft$$

$$t_{ftg} = 2\,ft$$

$$t_{wall} = 1\,ft$$

$$h_{wall} = 8\,ft$$

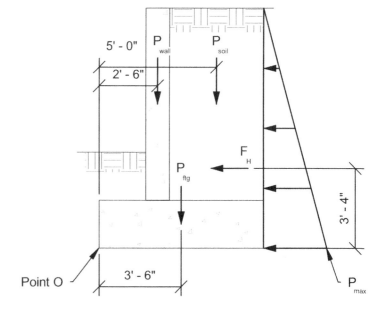

AM PRACTICE EXAM – Version B ANSWERS

$b_{toe} = 2 ft$ $b_{heel} = 4 ft$ $b_{ftg} = b_{toe} + b_{heel} + t_{wall} = 7 ft$

Calculating overturning force:

$$P_{max} = K_a \cdot \gamma_{soil} \cdot h = 0.33 \cdot \left(115 \frac{lb}{ft^3}\right) \cdot 10 ft = 379.5 \frac{lb}{ft^2}$$

$$F_H = \frac{1}{2} \cdot P_{max} \cdot h = \frac{1}{2} \cdot \left(379.5 \frac{lb}{ft^2}\right) \cdot 10 ft = 1897.5 \frac{lb}{ft}$$

Calculating overturning moment at toe (Point O):

$$M_{ot} = F_H \cdot \left(\frac{1}{3} \cdot h\right) = 1897.5 \frac{lb}{ft} \cdot \left(\frac{1}{3} \cdot 10 ft\right) = 6325 \frac{lb \cdot ft}{ft}$$

Calculating resisting forces:

$$P_{soil} = \gamma_{soil} \cdot b_{heel} \cdot h_{wall} = 3680 \frac{lb}{ft}$$

$$P_{wall} = \gamma_{conc} \cdot t_{wall} \cdot h_{wall} = 1200 \frac{lb}{ft}$$

$$P_{ftg} = \gamma_{conc} \cdot t_{ftg} \cdot b_{ftg} = 2100 \frac{lb}{ft}$$

Calculate resisting moment:

$$M_{resist} = (P_{soil} \cdot 5 ft) + (P_{wall} \cdot 2.5 ft) + (P_{ftg} \cdot 3.5 ft) = 28,750 \frac{lb \cdot ft}{ft}$$

Solve for factor of safety:

$$FS = \frac{M_{resist}}{M_{ot}} = \frac{28,750 \frac{lb \cdot ft}{ft}}{6,325 \frac{lb \cdot ft}{ft}} = 4.545$$

Answer: 4.55

AM PRACTICE EXAM – Version B ANSWERS

3. The maximum unfactored shear stress (psi) for the beam shown below is most nearly:

   (A) 50.00
   (B) 36.67
   (C) 33.33
   (D) 20.00

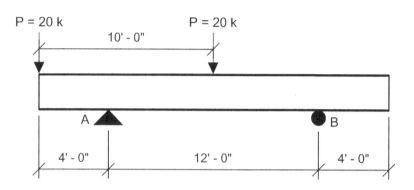
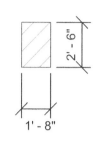

BEAM CROSS SECTION

Sum moments about point B = 0.

$$(20k \cdot 16\,ft) - (R_A \cdot 12\,ft) + (20k \cdot (16\,ft - 10\,ft)) = 0$$

$$R_A = \frac{(20k \cdot (16\,ft - 10\,ft)) + (20k \cdot 16\,ft)}{12\,ft} = 36.67k$$

Sum forces in y-direction = 0:

$$R_A + R_B - 20k - 20k = 0$$

$$R_B = 3.33k$$

Draw Shear Diagram →

Max shear force = 20k

Solve maximum shear stress:

$$\tau_{max} = \frac{3V}{2A} = \frac{3 \cdot 20k}{2 \cdot 20in \cdot 30in} = 0.05 \text{ ksi} = 50 \text{ psi}$$

Answer: 50.00

AM PRACTICE EXAM – Version B ANSWERS

4. A vertical excavation is to be performed in a deep clay soil. The clay has a unit weight and an unconfined compressive strength of 110 pcf and 2250 psf respectively. The maximum vertical cut (ft) without requiring a lateral support is most nearly:

   (A) 20.45
   (B) 25.20
   (C) 30.67
   (D) 40.91

Given:

$$\gamma = 110 \frac{lb}{ft^3}$$

$$q_u = 2250 \frac{lb}{ft^2}$$

Calculate cohesion of clay soil:

$$c = \frac{q_u}{2} = 1125 \frac{lb}{ft^2}$$

Solve for max allowable vertical cut:

$$K_a = 1.0 \text{ (for clay soils)}$$

$$z_{cr} = \frac{2 \cdot c}{\gamma \cdot \sqrt{K_a}} = \frac{2 \cdot \left(1125 \frac{lb}{ft^2}\right)}{\left(110 \frac{lb}{ft^3}\right) \cdot \sqrt{1.0}} = 20.45 \, ft$$

Answer: 20.45

AM PRACTICE EXAM – Version B ANSWERS

5. A normally consolidated clay soil is subjected to the surcharge shown below. The compression and recompression indexes are 0.10 and 0.05 respectively. The settlement (in) due to the surcharge is most nearly:

   (A) 2.8
   (B) 3.5
   (C) 3.7
   (D) 3.9

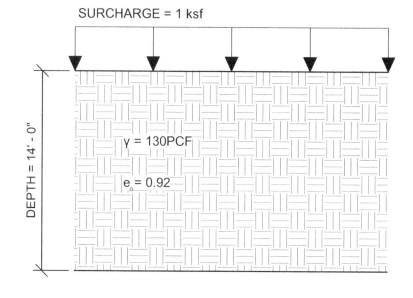

Given:

C_c = 0.10

C_r = 0.05

e_o = 0.92

H = 14 ft

Calculate initial effective overburden pressure at the midpoint of the soil layer:

$$\sigma'_0 = 7\,ft \cdot \left(130\,\frac{lb}{ft^3}\right) = 910\,\frac{lb}{ft^2}$$

Calculate final effective overburden pressure at the midpoint of the soil layer:

$$\sigma'_f = 7\,ft \cdot \left(130\,\frac{lb}{ft^3}\right) + 1000\,\frac{lb}{ft^2} = 1910\,\frac{lb}{ft^2}$$

Solve for primary consolidation:

$$S_c = \left(\frac{C_c}{1+e_o}\right) \cdot H \cdot \log\left(\frac{\sigma'_f}{\sigma'_o}\right)$$

$$S_c = \left(\frac{0.10}{1+0.92}\right) \cdot (14\,ft) \cdot \log\left(\frac{1910\,psf}{910\,psf}\right) = 0.235\,ft$$

$$S_c = 2.81\,in$$

Answer: 2.8

AM PRACTICE EXAM – Version B ANSWERS

6. A saturated soil specimen has the following properties:

   Weight before oven drying: 3.5 lb
   Weight after oven drying: 3.2 lb
   Specific gravity: 2.5

   The void ratio of the soil specimen is most nearly:

   (A) 0.15
   (B) 0.18
   (C) 0.20
   (D) 0.23

Given:

$W = 3.5 lb$ (total weight)

$W_s = 3.2 lb$ (weight of solids)

$G_s = 2.5$ (specific gravity)

$S = 1.0$ (S= 1.0 for saturated soils)

Calculate weight of water in the soil:

$$W_w = W - W_s = 3.5 lb - 3.2 lb = 0.3 lb$$

Calculate void ratio:

$$e = \frac{W_w \cdot G_s}{W_s \cdot S} = \frac{(0.3 lb) \cdot 2.5}{(3.2 lb) \cdot 1.0} = 0.23$$

Answer: 0.23

AM PRACTICE EXAM – Version B ANSWERS

7. A falling-head permeability test was performed on a clayey silt. The data collected is shown in the table below. The coefficient of permeability (cm/sec) is most nearly:

   (A) 1.911E-5
   (B) 1.712E-5
   (C) 1.655E-5
   (D) 1.655E-6

Experiment #1	
Area of specimen	85.00 cm²
Length of specimen	15.25 cm
Apparatus Standpipe Area	2.00 cm²
Water Height (start)	100 cm
Water Height (end)	88 cm
Test duration	40 min
Temperature	30°C

Solving for coefficient of permeability:

Note: The falling head permeability test is used for fine-grained soils.

The constant head permeability test is used for coarse-grained soils.

Equation for falling-head permeability test:

$$k = \frac{a \cdot L}{A \cdot t} \cdot \ln\left(\frac{h_o}{h_1}\right) = \frac{(2.00 cm) \cdot (15.25 cm)}{(85.00 cm) \cdot (40 min) \cdot \left(\frac{60 sec}{min}\right)} \cdot \ln\left(\frac{100 cm}{88 cm}\right)$$

$$k = 1.911 \cdot 10^{-5} \frac{cm}{sec}$$

Answer: 1.911E-5

AM PRACTICE EXAM – Version B ANSWERS

8. Which of the following should be added to the concrete mix to best protect against freeze-thaw cycles?

   (A) Retarding admixture
   (B) Accelerating admixture
   (C) Type III cement
   (D) Air-entraining admixture

An air-entraining admixture should be added to the mix to protect against freeze-thaw. Air-entraining admixtures provide air pockets in the concrete which allow the water to expand when frozen without distressing the concrete.

Answer: Air-entraining admixture

9. A roadway with a roughness coefficient of 0.014 is shown below. The longitudinal slope of the roadway is 2.0%. The minimum width of gutter (ft) required to contain a design flow of 3.0 cfs in its entirety is most nearly:

   (A) 5.0
   (B) 5.5
   (C) 6.5
   (D) 10.0

$Q = 3.0 \dfrac{ft^3}{sec}$

$n = 0.014$

$S_x = 0.035$ (slope of shoulder) $\qquad S = 0.02$ (longitudinal slope)

Solve for shoulder width (flow in open, triangular channel):

$$x = \dfrac{(1.79 \cdot Q \cdot n)^{\frac{3}{8}}}{S_x^{\frac{5}{8}} \cdot S^{\frac{3}{16}}}$$

$$x = \dfrac{\left(1.79 \cdot 3.0 \dfrac{ft^3}{sec} \cdot 0.014\right)^{\frac{3}{8}}}{0.035^{\frac{5}{8}} \cdot 0.02^{\frac{3}{16}}} = 6.412 \ ft$$

Answer: 6.5

AM PRACTICE EXAM – Version B ANSWERS

10. Traffic flow was observed along an arterial street. The vehicles were traveling an average of 55mph and were spaced 500 feet apart. The flow of traffic (veh/hr) is most nearly:

    (A) 240
    (B) 560
    (C) 580
    (D) 680

Given:

$$s = 500 \frac{ft}{veh}$$

$$v = 55 mph$$

Calculate the vehicle density:

$$k = \frac{1}{s} = 0.002 \frac{veh}{ft}$$

Solve for traffic flow:

$$q = k \cdot v = \left(0.002 \frac{veh}{ft}\right) \cdot \left(55 \frac{mi}{hr}\right) \cdot \left(\frac{5280 \, ft}{mi}\right) = 580.80 \frac{veh}{hr}$$

Answer: 580

AM PRACTICE EXAM – Version B ANSWERS

11. Which members from the truss below are zero-force members?

   (A) Member CG
   (B) Member CH and Member CF
   (C) Member BF and Member DH
   (D) Member BF, Member DH, Member CG

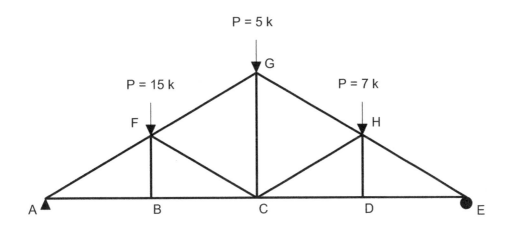

Criteria of zero-force members:

1) Two non-collinear members form a truss joint in which no external load is applied. Each of the members are zero force members.

2) Three members form a truss joint for which two of the members are collinear, AND there is no external load applied at that joint. The member which is non-collinear is a zero-force member.

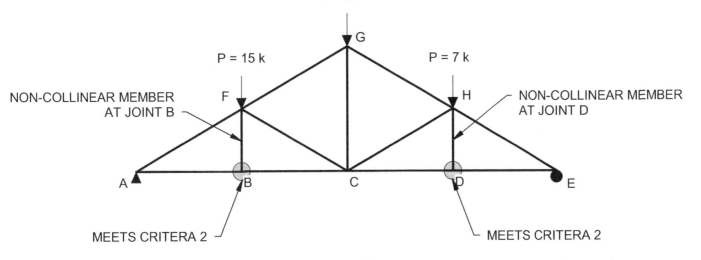

Answer: Member BF and Member DH

AM PRACTICE EXAM – Version B ANSWERS

12. A pressure drop of 12 psig is measured across a 150 ft long ductile iron pipe. The water velocity (ft/sec) in the pipe is most nearly:
Note: the change in elevation and velocity heads are negligible.

Pipe Properties:
Inside Diameter = 6 in
friction coefficient = 140

(A) 20
(B) 18
(C) 17
(D) 15

Given:

$\Delta P = 12\,psi \qquad C = 140 \qquad D = 6in = 0.5\,ft \qquad L = 150\,ft$

Calculate friction loss/head:

$\Delta P = \gamma \cdot h_f$ or

$$h_f = \frac{\Delta P}{\gamma} = \frac{12\,\frac{lb}{in^2} \cdot \left(\frac{144\,in^2}{ft^2}\right)}{62.4\,\frac{lb}{ft^3}}$$

$h_f = 27.7\,ft$

Using Hazen-Williams equation (units of ft and sec):

$$h_f = \frac{3.022 \cdot v^{1.85} \cdot L}{C^{1.85} \cdot D^{1.17}}$$

Rewriting and solving for v:

$$v = \left(\frac{h_f \cdot C^{1.85} \cdot D^{1.17}}{3.022 \cdot L}\right)^{0.541}$$

$v = 19.98\,ft/\sec$

Answer: 20

AM PRACTICE EXAM – Version B ANSWERS

13. The flow through the given open channel is most nearly characterized by:

Given:
V= 10ft/sec

(A) Critical flow
(B) Supercritical flow
(C) Subcritical flow
(D) None of the above

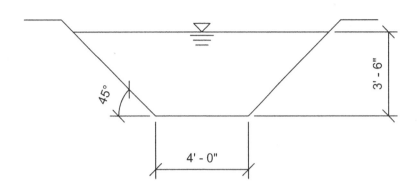

Determine the Froude Number to characterize flow:

$$Fr = \frac{v}{\sqrt{g \cdot D_h}}$$

Calculate mean hydraulic depth:

$$D_h = \frac{Area\ of\ flow}{width\ of\ open\ surface}$$

$$Area_{flow} = 3.5\ ft \left[\frac{4\ ft + (4\ ft + 3.5\ ft + 3.5\ ft)}{2} \right] = 26.25\ ft^2$$

$$width_{open} = 4\ ft + 3.5\ ft + 3.5\ ft = 11\ ft$$

$$D_h = \frac{26.25\ ft^2}{11\ ft} = 2.386\ ft$$

$$Fr = \frac{10\ \frac{ft}{sec}}{\sqrt{\left(32.2\ \frac{ft}{sec^2}\right) \cdot 2.386\ ft}} = 1.14$$

Froude Number is greater than 1; therefore, flow is supercritical.

Answer: Supercritical flow

Copyright 2018 by CivilPEPractice

AM PRACTICE EXAM – Version B ANSWERS

14. A working pressure of 45 psig is needed at Point B of an 8-inch (I.D.) steel pipe. The velocity and friction factor are 6.0 ft/s and 0.0165, respectively. The pressure head (ft) at Point A is most nearly:

Note: Minor losses are negligible.

(A) 190
(B) 175
(C) 170
(D) 160

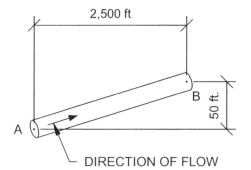

Given:

$v = 6.0 \dfrac{ft}{\sec} \qquad D = 8in$

$f = 0.0165 \qquad P_B = 45\,psi$

$L = \sqrt{(2500ft)^2 + (50ft)^2} = 2500.50\,ft$

Set up energy balance equation:

$H_A = H_B + h_f$

$h_{p_A} + h_{v_A} + h_{z_A} = h_{p_B} + h_{v_B} + h_{z_B} + h_f$

Rewriting for pressure head at point A:

Note: velocity terms are equal and can be ignored.

$h_{p_A} = h_{p_B} + h_{v_B} + h_{z_B} + h_f - h_{v_A} - h_{z_A}$

Solving for other terms:

Converting pressure at B to head:

$h_{p_B} = 45 \dfrac{lb}{in^2} \cdot \left(\dfrac{ft^3}{62.4\,lb}\right) \cdot \left(\dfrac{144\,in^2}{ft^2}\right) = 103.85\,ft$

Elevation head at points A and B

$h_{zA} = 0\,ft$

$h_{zB} = 50\,ft$

AM PRACTICE EXAM – Version B ANSWERS

friction head loss:

$$h_f = \frac{f \cdot L \cdot v^2}{2 \cdot D \cdot g} = \frac{0.0165 \cdot (2500.50\,ft) \cdot \left(6.0\,\frac{ft}{\sec}\right)^2}{2 \cdot (8in) \cdot \left(\frac{ft}{12in}\right) \cdot \left(32.2\,\frac{ft}{\sec^2}\right)} = 34.595\,ft$$

Solve for pressure head at point A

$$h_{P_A} = h_{P_B} + h_{z_B} + h_f - h_{z_A} = 103.85\,ft + 50\,ft + 34.595\,ft - 0\,ft = 188.44\,ft$$

Answer: 190

15. A -2.75% grade intersects a +1.50% grade at Sta. 20+00 and elevation 200.00 ft. A 300-ft vertical curve connects the two grades. The elevation (feet) of the turning point is most nearly:

(A) **201.46**
(B) 201.14
(C) 200.14
(D) 200.00

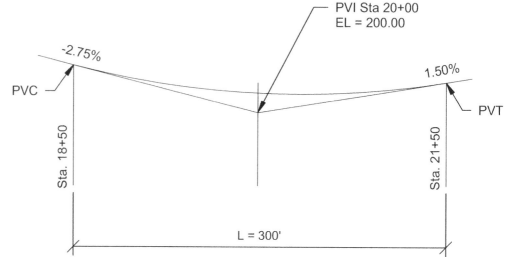

Given:

$G_1 = -0.0275$

$G_2 = 0.015$

$L = 300 \ ft$

$PVI_{EL} = 200.00 \ ft$

Solve turning point elevation:

$$R = \frac{(G_2 - G_1)}{L} = \frac{(0.015 - (-0.0275))}{300 \ ft} = \frac{1.417 \cdot 10^{-4}}{ft}$$

$$x = \frac{-G_1}{R} = \frac{0.0275}{1.417 \cdot 10^{-4} \cdot \left(\frac{1}{ft}\right)} = 194.07 \ ft$$

$$PVC_{EL} = PVI_{EL} - G_1 \cdot \left(\frac{L}{2}\right) = 200.00 \ ft - \left(-0.0275 \cdot \frac{300 \ ft}{2}\right) = 204.125 \ ft$$

$$TP_{EL} = \frac{R}{2} \cdot x^2 + G_1 \cdot x + PVC_{EL}$$

$$TP_{EL} = \frac{1.417 \cdot 10^{-4}}{2} \cdot (194.07 \ ft)^2 - 0.0275 \cdot 194.07 \ ft + 204.125 \ ft = 201.46 \ ft$$

Answer 201.46

AM PRACTICE EXAM – Version B ANSWERS

16. A 5.0 acre concrete parking lot is to be constructed. Which of the following compactors is best suited to compact and finish the base layer comprised of dense graded aggregate?

   (A) Sheepsfoot Roller
   (B) Tamping Rollers
   (C) Vibrating Smooth Wheel Roller
   (D) Pneumatic Tire Roller

   - Sheepsfoot Rollers are best for compacting fine-grained soils.
   - Similar to sheepsfoot rollers, tamping rollers are also best used on fine-grained soils.
   - Vibrating smooth wheel rollers are ideal for upper soil layers and coarse-grained soils such as dense graded aggregate.
   - Pneumatic tire rollers are best used on coarse-grained soils with fines or asphalt pavements.

   ## Answer: Vibrating Smooth Wheel Roller

17. The framing for a mechanical mezzanine is shown below. The top of framing is 15'-0" above the foundation. The weight of the mezzanine framing (tons) is most nearly:

   (A) 6.5
   (B) 9.0
   (C) 13.1
   (D) 18.0

   Note: include weight of columns and beams.

   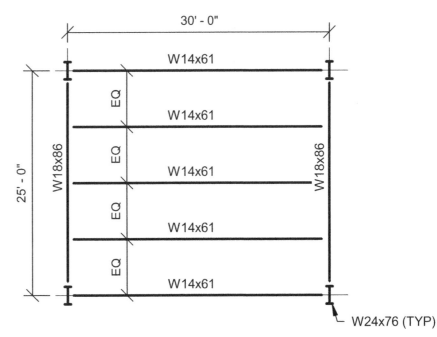

   Calculating steel weight:

   $$w_{tot} = \left(5 \cdot \frac{61\,lb}{ft} \cdot 30\,ft\right) + \left(2 \cdot \frac{86\,lb}{ft} \cdot 25\,ft\right) + \left(4 \cdot \frac{76\,lb}{ft} \cdot 15\,ft\right) = 18{,}010\,lb$$

   $$w_{tot} = 9.005\,tons$$

   ## Answer: 9.0

AM PRACTICE EXAM – Version B ANSWERS

18. An investor purchases a commercial building in a large city adjacent to the main river. The investor takes out a 20-year mortgage to finance the building. The probability of at least one 100-year flood occurring during the mortgage period is most nearly:

    (A) 0.99
    (B) 0.74
    (C) 0.30
    (D) 0.18

Given: $F = 100$ (frequency) $n = 20$ (periods)

Calculate probability:

$$p = 1 - \left(1 - \frac{1}{F}\right)^n \rightarrow p = 1 - \left(1 - \frac{1}{100}\right)^{20} = 0.182$$

Answer: 0.18

19. A circular pipe is to be installed as part of a residential sewer system. The pipe will be installed on a 2% grade and has a roughness coefficient of 0.012. The minimum full-flow capacity shall be 3.2 ft³/sec. The minimum inside diameter of pipe (in) that can meet this requirement is most nearly:

    (A) 8
    (B) 10
    (C) 12
    (D) 18

Given: $n = 0.012$ $S = 0.02$ $Q_{full} = 3.2 \dfrac{ft^3}{sec}$

Solve for required pipe diameter (circular pipe flowing full):

$$D = 1.335 \cdot \left(\frac{Q_{full} \cdot n}{\sqrt{S}}\right)^{\frac{3}{8}} \rightarrow D = 1.335 \cdot \left(\frac{3.2 \frac{ft^3}{sec} \cdot 0.012}{\sqrt{0.02}}\right)^{\frac{3}{8}} = 0.819 \, ft = 9.825 \, in$$

Answer: 10

Copyright 2018 by CivilPEPractice

AM PRACTICE EXAM – Version B ANSWERS

20. The maximum unfactored bending stress (ksi) of the W24x76 beam shown below is most nearly:

    Note: Neglect self-weight of beam. All loads given are unfactored.

    (A) 12
    (B) 16
    (C) 20
    (D) 25

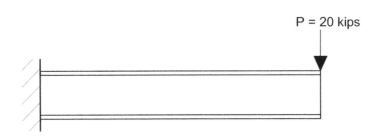

Given:

$P = 20k$

$L = 12ft$

$S_x = 176in^3$

$S_y = 18.4in^3$

Calculate the maximum moment:

$M = P \cdot L = 20k \cdot 12\,ft = 240k \cdot ft$

Calculate the bending stress:

$$f_b = \frac{M}{S_x}$$

Note: use S_x since load is acting about the beam's strong axis

$$f_b = \frac{240k \cdot ft \cdot \left(\frac{12in}{ft}\right)}{176in^3} = \frac{16.36k}{in^2}$$

Answer: 16

Copyright 2018 by CivilPEPractice
For more problems visit https://civilpepractice.com/

AM PRACTICE EXAM – Version B ANSWERS

21. The unfactored axial stress (ksi) in column A is most nearly:

   (A) 0.20
   (B) 1.50
   (C) 4.25
   (D) 5.25

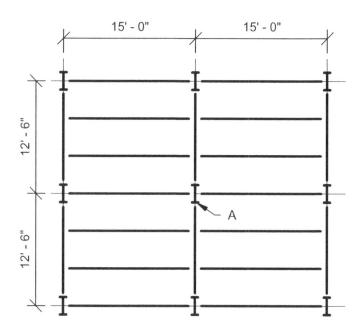

Loads:
DL = 40 psf (includes self-weight)
LL = 100 psf

Column properties:
d = 12 in
b_f = 12 in
A = 5 in²

Calculate tributary area of column:

$$A_{trib} = \left(\frac{15\,ft + 15\,ft}{2}\right) \cdot \left(\frac{12.5\,ft + 12.5\,ft}{2}\right) = 187.5\,ft^2$$

Calculate load on column:

$$P = (DL + LL) \cdot A_{trib} = \left(\frac{40\,lb}{ft^2} + \frac{100\,lb}{ft^2}\right) \cdot 187.5\,ft^2 = 26{,}250\,lb = 26.25\,k$$

Calculate compressive stress:

$$\sigma = \frac{P}{A_{col}} = \frac{26.25k}{5in^2} = 5.25\,\frac{k}{in^2}$$

Answer: 5.25

Copyright 2018 by CivilPEPractice
For more problems visit https://civilpepractice.com/

AM PRACTICE EXAM – Version B ANSWERS

22. The length of curve AB (ft) measured along its arc is most nearly:

   (A) 698
   (B) 705
   (C) 812
   (D) 819

Given:

R = 800ft   I = 50°30'00" = 50.5°

Solving Length of curve:

$$L_{AB} = 2 \cdot \pi \cdot R \cdot \left(\frac{I}{360}\right)$$

$$L_{AB} = 2 \cdot \pi \cdot 800 ft \cdot \left(\frac{50.5}{360}\right) = 705.11 ft$$

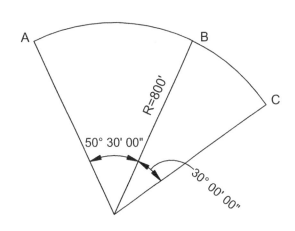

Answer: 705

23. Which of the following concrete beams offer the greatest shear resistance to load P?

   (A) A
   (B) B
   (C) C
   (D) D

Beam "C" has a similar shear area relative to the others, but has greater resistance due to the contribution of shear reinforcement.

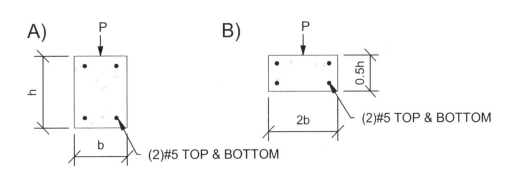

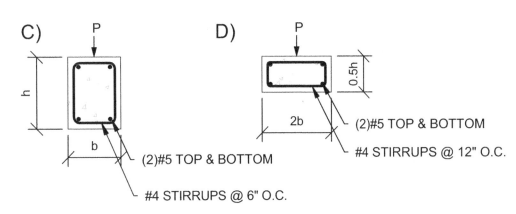

Answer: C

AM PRACTICE EXAM – Version B ANSWERS

24. The survey crew gathers information at a future construction site. A summary of the data collected is given below. The ground elevation at point A is most nearly:

   (A) 93.21
   (B) 94.12
   (C) 95.55
   (D) 97.51

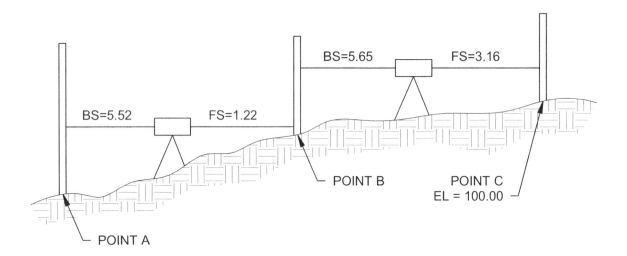

Calculate ground elevation at point B:

$$EL_B = EL_C + FS_C - BS_B = 100.00 + 3.16 - 5.65 = 97.51$$

Solve for ground elevation at point A:

$$EL_A = EL_B + FS_B - BS_A = 97.51 + 1.22 - 5.52 = 93.21$$

Answer: 93.21

25. Total deflection for the beam shown below shall not exceed L/360 or 0.5 in. What is the lightest beam that will satisfy the deflection requirements?

Note: Neglect self-weight of beam. E = 29,000 ksi

(A) W18x65
(B) W18x71
(C) W18x76
(D) W18X86

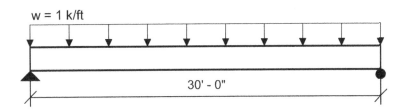

Given:

$$w = 1 \frac{k}{ft}$$

$$E = 29,000 \frac{kip}{in^2}$$

$$L = 30 ft$$

W18x65: $I_x = 1070 in^4$
W18x71: $I_x = 1170 in^4$
W18x76: $I_x = 1330 in^4$
W18X86: $I_x = 1530 in^4$

Calculate the controlling deflection criteria:

$$\frac{L}{360} = \frac{30 ft}{360} = 0.083 ft = 1 in$$

$$\frac{L}{360} > 0.5 in, \therefore 0.5 in \text{ controls}$$

Solve for required moment of inertia:

$$\Delta = \frac{5 \cdot w \cdot L^4}{384 \cdot E \cdot I_x} \xrightarrow{solving\ for\ I_x} I_x = \frac{5 \cdot w \cdot L^4}{384 \cdot E \cdot \Delta}$$

$$I_x = \frac{5 \cdot \left(1 \frac{k}{ft}\right) \cdot (30 ft)^4}{384 \cdot \left(29,000 \frac{k}{in^2}\right) \cdot (0.5 in)} \cdot \left(\frac{1728 in^3}{ft^3}\right) = 1257 in^4$$

The W18x76 is the lightest member to meet the deflection criteria.

Answer: W18x76

AM PRACTICE EXAM – Version B ANSWERS

26. A rectangular wastewater tank has dimensions of 40 ft x 60 ft x 15 ft deep and is receiving a flow of 600 gal/min. The temperature and unit weight of the wastewater are 70°F and 65pcf respectively. The hydraulic detention time (hr) for the tank is most nearly:

   (A) 7.12
   (B) 7.48
   (C) 7.65
   (D) 7.76

Given:

$$Q = 600 \frac{gal}{min}$$

B = 40ft

L = 60ft

H = 15ft

$$V = (40\ ft) \cdot (60\ ft) \cdot (15\ ft) = 36,000\ ft^3$$

Solve for detention time:

$$t_d = \frac{V}{Q} = \frac{36,000\ ft^3}{\left(600 \frac{gal}{min}\right) \cdot \left(\frac{ft^3}{7.48 gal}\right) \cdot \left(\frac{60\ min}{hr}\right)}$$

$$t_d = 7.48\ hr$$

Answer: 7.48

AM PRACTICE EXAM – Version B ANSWERS

27. A soil profile is given below. The difference in total vertical stresses (psf) at Point A and Point B is most nearly:

    (A) 2850
    (B) 2784
    (C) 2551
    (D) 1903

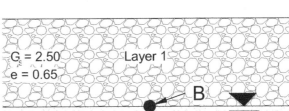

Given:

$e_1 = 0.65 \qquad e_2 = 0.50$

$G_{s1} = 2.50 \qquad G_{s2} = 2.55$

$h = 10\,ft \qquad h = 15\,ft$

Calculate applicable soil unit weights:

$$\gamma_{d1} = \frac{G_{s1} \cdot \gamma_{water}}{1 + e_1} = \frac{2.50 \cdot 62.4 \,\frac{lb}{ft^3}}{1 + 0.65} = 94.545 \,\frac{lb}{ft^3}$$

$$\gamma_{sat2} = \frac{(G_{s2} + e_2) \cdot \gamma_{water}}{1 + e_2} = \frac{(2.55 + 0.50) \cdot 62.4 \,\frac{lb}{ft^3}}{1 + 0.50} = 126.88 \,\frac{lb}{ft^3}$$

Calculate total vertical stress at Point A:

$$\sigma_A = \gamma_{d1} \cdot h_1 + \gamma_{sat2} \cdot h_2 = 94.545 \,\frac{lb}{ft^3} \cdot 10\,ft + 126.88 \,\frac{lb}{ft^3} \cdot 15\,ft = 2848.655 \,\frac{lb}{ft^2}$$

Calculate total vertical stress at Point B:

$$\sigma_B = \gamma_{d1} \cdot h_1 = 94.545 \,\frac{lb}{ft^3} \cdot 10\,ft = 945.455 \,\frac{lb}{ft^2}$$

Calculate difference in total vertical stress.

$$\sigma_A - \sigma_B = 1903.200 \,\frac{lb}{ft^2}$$

Answer: 1903

Copyright 2018 by CivilPEPractice
For more problems visit https://civilpepractice.com/

AM PRACTICE EXAM – Version B ANSWERS

28. A 20-ft thick clay soil layer is confined between two sand layers. The clay has a coefficient of consolidation of 0.28 ft²/day. The time (days) when 50% of the total clay settlement will occur is most nearly:

   (A) 25
   (B) 70
   (C) 160
   (D) 281

U	Tv
0.10	0.008
0.20	0.031
0.30	0.071
0.40	0.126
0.50	0.197
0.55	0.238
0.60	0.287
0.65	0.340
0.70	0.403
0.75	0.477
0.80	0.567
0.85	0.684
0.90	0.848
0.95	1.129
0.99	1.781

U = degree of consolidation
T_v = time factor

Given:

$$c_v = 0.28 \frac{ft^2}{day}$$

$$H_d = 10 \, ft \; (taken \; at \; midpoint \; of \; layer)$$

Note: Drainage height equals half the height of clay layer since there is drainage through the upper and lower sand layers (two-way drainage).

$U = 0.50$

$T_v = 0.197$ (from table, $U = 0.50$)

Solve for time:

$$T_v = \frac{c_v \cdot t}{H_d^2}$$

Rewriting and solving for t:

$$t = \frac{T_v \cdot H_d^2}{c_v} = \frac{0.197 \cdot (10 \, ft)^2}{0.28 \frac{ft^2}{day}} = 70.36 \, days$$

Answer: 70

AM PRACTICE EXAM – Version B ANSWERS

29. A concrete wall retains 16' of sand as shown below. The at-rest resultant (lb/ft) induced on the wall due to lateral earth pressure is most nearly:

   (A) 1,499
   (B) 10,550
   (C) 11,994
   (D) 12,202

   $\gamma_{sat} = 125 \dfrac{lb}{ft^3}$

   $\varphi' = 30°$

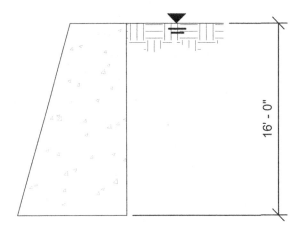

Given:

$\gamma_{sat} = 125 \dfrac{lb}{ft^3}$

$\phi' = 30$

H = 16ft

$\gamma_{water} = 62.4 \dfrac{lb}{ft^3}$

Calculate at rest pressure coefficient

$K_o = 1 - \sin(\phi') = 0.50$

Calculate effective unit weight of soil:

$\gamma' = \gamma_{sat} - \gamma_{water} = 125 \dfrac{lb}{ft^3} - 62.4 \dfrac{lb}{ft^3} = 62.6 \dfrac{lb}{ft^3}$

Calculate at rest pressure:

$p_0 = K_0 \cdot (\gamma' \cdot H) + \gamma_{water} \cdot H = 0.50 \cdot \left(62.6 \dfrac{lb}{ft^3} \cdot 16 ft\right) + 62.4 \dfrac{lb}{ft^3} \cdot 16 ft$

$p_0 = 1499.2 \dfrac{lb}{ft^2}$

Calculate resultant force:

$R_0 = \dfrac{1}{2} \cdot p_0 \cdot H = \dfrac{1}{2} \cdot \left(1499.2 \dfrac{lb}{ft^2}\right) \cdot 16 ft = 11{,}993.6 \dfrac{lb}{ft}$

Answer: 11,994

AM PRACTICE EXAM – Version B ANSWERS

30. According to OSHA regulations which of the following is an acceptable means of fall protection for an employee working on a roof structure 20ft above the ground level?

    (A) Guardrail systems
    (B) Safety net systems
    (C) Fall protection is not required
    (D) Guardrail systems and safety net systems

According to OSHA Construction Industry Regulations Part 1926.760(a) the following fall protection systems are acceptable:

- guardrail systems
- safety net systems
- personal fall arrest systems
- positioning device systems
- fall restraint systems

Answer: guardrail systems and safety net systems

AM PRACTICE EXAM – Version B ANSWERS

31. The total vertical stress (psf) at Point A of the soil profile shown is most nearly:

   (A) 850
   (B) 1350
   (C) 1750
   (D) 2600

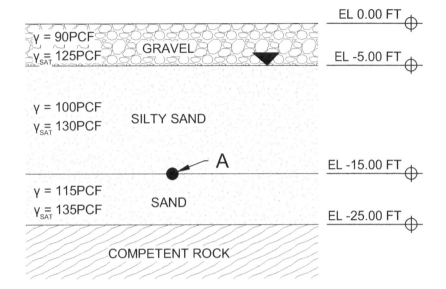

Calculate total stress at point A: $\sigma_v = \gamma_{GRAVEL} \cdot H_1 + \gamma_{sat-siltsand} \cdot H_2$

$$\sigma_v = \left(90 \frac{lb}{ft^3} \cdot 5\,ft\right) + \left(130 \frac{lb}{ft^3} \cdot 10\,ft\right) = 1750 \frac{lb}{ft^2}$$

Answer: 1750

32. Concrete creep is best defined as:

   (A) Weakening caused by repeatedly applied loads.
   (B) Contracting of a hardened concrete mixture due to the loss of capillary water.
   (C) Deterioration of concrete due to freeze thaw cycles.
   (D) Time-dependent deformation under a sustained load.

A) Fatigue is the weakening caused by repeatedly applied loads.
B) Concrete shrinkage is the contracting of a hardened concrete mixture due to the loss of capillary water.
C) Describes the potential damages of concrete when exposed to freeze/thaw conditions.
D) Creep is the time-dependent deformation under a sustained load.

Answer: D

AM PRACTICE EXAM – Version B ANSWERS

33. The net excavation (yd³) from Station 500+00 to 507+00 is most nearly:

   (A) 1400 (cut)
   (B) 2000 (cut)
   (C) 1400 (fill)
   (D) 2000 (fill)

Determine the average cut/fill between each station given:

Station	End Area (ft²) Cut	End Area (ft²) Fill	Length	Cut (ft³)	Fill (ft³)
500 + 00	0	325			
			100	0	27,500
501 + 00	0	225			
			100	0	26,250
502 + 00	0	300			
			100	0	15,000
503 + 00	0	0			
			100	6,250	0
504 + 00	125	0			
			100	28,750	0
505 + 00	450	0			
			100	41,250	0
506 + 00	375	0			
			100	31,250	0
507 + 00	250	0			
				107,500	68,750

$Cut_{total} = 107,500 \, ft^3$

$Fill_{total} = 68,750 \, ft^3$

Solve for net excavation:

$Net = Cut_{total} - Fill_{total} = 38,750 \, ft^3$

$Net = 1435 \cdot yd^3 \; (Cut)$

Answer: 1400 (Cut)

AM PRACTICE EXAM – Version B ANSWERS

34. Concrete forms shall be designed for which of the following loads:

    (A) Weight of concrete and Construction Live Loads
    (B) Hydrostatic pressure of wet concrete
    (C) Both A and B
    (D) None of the Above

    According to ACI 347 forms shall be designed for vertical loads (includes dead and live loads) and lateral pressure of concrete.

Answer: (C) Both A and B

AM PRACTICE EXAM – Version B ANSWERS

35. A contractor is required to install 12 bollards outside of a new fire station. It takes 2 laborers 2 hours to install one bollard. The pay rate for a single laborer is $25/hr. The total cost to install all 12 bollards including labor and material is most nearly:

(A) $375
(B) $1200
(C) $3300
(D) $4500

Material	Unit	Unit Cost
Steel Pipe	LF	$25
Concrete Foundation	CY	$200
Concrete Fill Inside Pipe	CY	$150

Determine the material cost for a single bollard:

Given:

$$d_{fdn} = 2.5\,ft \qquad d_{pipe} = 6in$$

$$t_{pipe} = 0.25in$$

$$Cost_{mat} = \left(\frac{\$25}{ft} \cdot 6.5\,ft\right) + \left(\frac{\$200}{yd^3} \cdot \frac{\pi \cdot d_{fdn}^2}{4} \cdot 3\,ft\right) + \left[\frac{\$150}{yd^3} \cdot \frac{\pi \cdot (d_{pipe} - 2 \cdot t_{pipe})^2}{4} \cdot 4\,ft\right]$$

$$Cost_{mat} = \$275$$

Determine the labor cost for a single bollard:

$$Cost_{lab} = 2 \cdot 2hr \cdot \frac{\$25}{hr} = \$100$$

Solve for total cost:

$$Cost_{total} = 12 \cdot (Cost_{mat} + Cost_{lab}) = \$4500$$

Answer: $4500

AM PRACTICE EXAM – Version B ANSWERS

36. A partial plan of the formwork for a suspended 8" thick concrete slab is given below. The girders are spaced at 8'-0" on center and the joists are spaced at 2'-0" on center. The superimposed live load on the formwork is 50psf. The maximum unfactored bending moment (lb-ft) in the formwork joist is most nearly:

Note: ignore self-weight of the joists

(A) 1215
(B) 1615
(C) 2430
(D) 3230

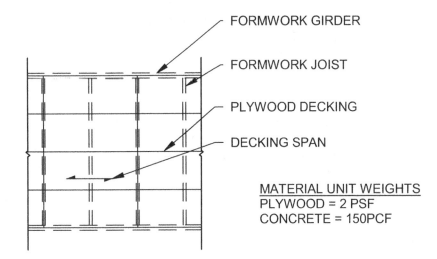

Given:

$b_{trib} = 2\,ft$ (trib width of joist)

$L = 8\,ft$ (length of joist)

$\gamma_{conc} = 150 \dfrac{lb}{ft^3}$ $w_{deck} = 2 \dfrac{lb}{ft^2}$ $t_{conc} = 8\,in$

Determine maximum uniform load on the joist:

$w_{DL} = (w_{deck} + \gamma_{conc} \cdot t_{conc}) \cdot b_{trib}$

$w_{DL} = (2 \dfrac{lb}{ft^2} + 150 \dfrac{lb}{ft^3} \cdot 0.67\,ft) \cdot 2\,ft = 204 \cdot \dfrac{lb}{ft}$

$w_{LL} = 50 \dfrac{lb}{ft^2} \cdot b_{trib} = 100 \cdot \dfrac{lb}{ft}$

$w_{tot} = w_{DL} + w_{LL} = 304 \cdot \dfrac{lb}{ft}$

Solve for maximum bending moment
Max moment for a simply supported beam subjected to a uniform load:

$M_{max} = \dfrac{W_{tot} \cdot L^2}{8} = 2432\,lb \cdot ft$

Answer: 2430

AM PRACTICE EXAM – Version B ANSWERS

37. A contractor is to construct a building addition for a shipping and receiving center. The foundation for the addition is to be constructed 5 feet above the existing building foundation. The new addition is to match the existing construction with shallow spread footings and concrete stem walls. Which of the following construction methods should be used to prevent overloading the existing footing?

   (A) Dowel the new foundation into the existing
   (B) Construct the new foundation outside the zone of influence of the existing footing
   (C) Both A and B
   (D) None of the Above

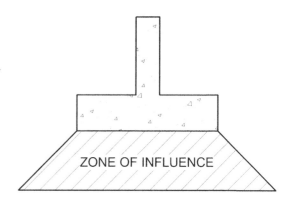

The influence zone is the zone of soil that is significantly influenced by an applied surface load. By constructing the new addition outside of this zone the contractor is reducing/eliminating the possibility of surcharging the existing foundation.

Answer: B

38. Which of the following is utilized as an erosion control method?

   (A) Geotextiles
   (B) Straw mulch
   (C) Hydraulic mulch
   (D) All of the Above

Geotextiles, straw mulch and hydraulic mulch are all used to prevent/minimize erosion.

Answer: All of the above

AM PRACTICE EXAM – Version B ANSWERS

39. A critical path diagram is shown below for an industrial construction project. The minimum number of days required to complete the project is most nearly:

   (A) 22
   (B) 23
   (C) 25
   (D) 27

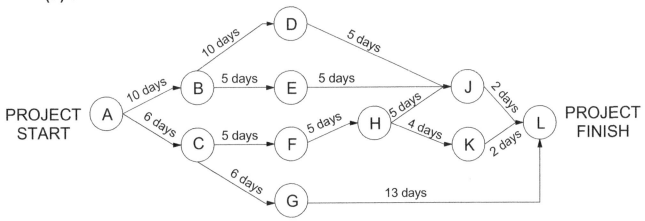

The critical path of the project A→B→D→J→L = 10 + 10 + 5 + 2 = 27 days

Answer: 27

40. An excavator with a bucket capacity of 3-yd³ has a standard operating cycle time of 55 seconds. One 24 yd³ dump truck is used to dispose of the excavated soil generated by the excavator. The total number of trips required to haul all of the excavated soil generated by one excavator during an 8 hour workday is most nearly:

   (A) 60
   (B) 65
   (C) 66
   (D) 68

Determine the daily production of the excavator:

$$V_{exc} = 3\, yd^3 \qquad t = 8\, hr \qquad t_{cycle} = 55\, sec$$

$$P = \frac{V_{exc}}{t_{cycle}} \cdot t = \frac{3\, yd^3}{55\, sec} \cdot 8\, hr \cdot \left(\frac{3600\, sec}{hr}\right) = 1570.91\, yd^3$$

Solve for number of trips:

$$V_{truck} = 24\, yd^3 \qquad Trips = \frac{Prod}{V_{truck}} = \frac{1570.91\, yd^3}{24\, yd^3} = 65.45$$

Answer: 66

Copyright 2018 by CivilPEPractice
For more problems visit https://civilpepractice.com/

Made in the USA
Monee, IL
09 June 2024

59521335R00070